Klaus-Dieter Sedlacek

Immortal consciousness

Author

The author Klaus-Dieter Sedlacek, born in 1948, studied mathematics and computer science. He completed his studies in 1975 with a diploma in mathematics. After a few years of professional experience, he founded his own company, which dealt with the development of application software. This he led for more than twenty-five years. As a mathematician, he is predestined to uncover and logically explain complex interrelationships in our world. Besides non-fiction books he also writes exciting novels.

Klaus-Dieter Sedlacek

Immortal
Consciousness

Space-time Phenomena
Evidence And Visions

Bibliografische Information der Deutschen Nationalbibliothek: Die Deutsche Nationalbibliothek verzeichnet diese Publikation in der Deutschen Nationalbibliografie; detaillierte bibliografische Daten sind im Internet über dnb.dnb.de abrufbar.

© 2020 Klaus-Dieter Sedlacek
https://toppbook.de

Cover composition: Sedlacek; basic image: Wolfgang Beyer, GNU license (see appendix).

Production and publishing
Herstellung und Verlag: BoD - Books on Demand, Norderstedt

ISBN: 978-3-7504-6212-0

Table of contents

Preface 7

Credibility of statements about reality 9

What comes after physics 35

A physical theory from the hereafter 59

Outside space and time 66

The primary consciousness of the vacuum 82

The true face of reality 111

Answers to basic questions of our being 123

    Free will 123
    The demiurge (world builder) and the answer to the
    question of meaning 125
    Immortality 130

Glossary 133

References 136

List of figures 138

Keyword index 140

Appendix 143

    GNU Free Documentation License 143

Preface

This book is neither about faith nor esotericism, but about evidence. Credible, scientific evidence, packed in a form that is understandable and comprehensible to anyone interested. The form of presentation is a frame story in which the fictional Professor Allman holds a course for his colleagues. Little by little, Professor Allman develops a robust scientific theory.

It is unusual for a scientific work to be structured like a non-fiction book and use a framework story. But this work also has an unusual content that concerns us all. It should not and must not gather dust in the libraries of the professional world, but is pressing for access to a broad public. Probably for the first time, it is possible to prove that consciousness exists outside the brain. This has hardly foreseeable consequences for our world view. Some of these consequences are presented.

The framework plot and the names of the course participants are fictitious, but the content put up for discussion is real. The unexplained phenomena presented here, which can be explained, have usually been known to experts for decades. However, because the phenomena have so far resisted any deeper explanation, scientists have not been able to present them to a wider audience in an understandable way. Science accepted them as inexplicable, dealt with them and got used to them until they seemed quite ordinary and natural. The larger connection behind the phenomena remained hidden.

Since I began to deal with it in my student days, the phenomena have not left me in peace. Again and again I thought that there must be an explanation for this. Unfortunately, during my professional life as a mathematician and computer scientist I did not have the time to look for a solution. Only now in my retirement did I get these.

To benefit from a deeper understanding of the content, it should not simply be consumed. Rather, it is necessary to think about the evidence in order to understand it. At this point I would like to quote a sentence by Immanuel Kant, who is one of the most important philosophers of modern times: *„Habe Mut, dich deines eigenen Verstandes zu bedienen!" (Berlinische Monatsschrift, 1784,2, S. 481–494).* The reward for the effort is an incredible, and for many people even gratifying, insight, of which I do not want to reveal too much at this point. I believe and hope that I have succeeded in explaining the phenomena and presenting the larger context in such a way that school knowledge is sufficient to grasp the fantastic and yet real consequences for the world and for you personally.

Klaus-Dieter Sedlacek

Credibility of statements about reality

> *This is where I get a foothold!*
> *Here they are realities,*
> *From here the spirit*
> *may argue with spirits,*
> *The double kingdom, the great one, is*
> *preparing itself.*

Johann Wolfgang von Goethe, Faust II

Monday, June 2

"What is reality? - What is the essence of spirit and matter? - Does man have free will? - What is the meaning of life? - Will our individual consciousness survive personal death - these are just a few of the issues we will be addressing in the coming days!

With his strong, deep voice, Professor Allman opens the well-attended course with the title: "On the immortality of consciousness - a new metaphysical world view".

Previously, the well-known physicist from Quantum City welcomed his interested colleagues from all departments.

The sun floods through the large windows of the secluded seminar room in the outpost of the AlbertEinstein University. The course building sits alone on top of a dune with a view over the vastness of the Atlantic Ocean and is often used when a small group of highly qualified specialists wants to enter new scientific territory and combine the wonders of naturalscience into one overall view. But the group that found its way to the seclusion, a good 150 km from Quantum City, this Monday, is by no means small. Despite the limitation to a maximum of 25 participants, another ten colleagues of the extravagant physics professor came from disciplines not specifically addressed. These are colleagues who did not want to miss the interesting topic under any circumstances. Professor Allman did not have the heart to reject her and so the classroom is packed.

The series of courses will last the whole week. Those who do not want to drive home on the evening of a seminar day can stay overnight in the affiliated hotel and combine the acquisition of knowledge with a few hours of holiday feeling.

The 43-year-old and 1.80 m tall Professor Allman strokes with his left hand over his full beard, which is only a few millimetres long, and looks around expectantly to see if there is already a reaction to his first sets. He looks likeable with his rounded face and the lively, friendly eyes flashing through his glasses. His brown checkered jacket and a blue silk scarf give him the extravagant look of an artist rather than a dry naturalscientist.

Courses for scientists of equal rank are characterized by a high degree of interactivity between the moderator and the course participants. So the first interposed question is not missing. Dr. Helena Anaximenes, a highly qualified mathematician from the front row, despite her red hair, smiling mischievously throws in: "Professor Allman, aren't you poaching your subject from the philosophers? What does this topic have to do with your department of physics?"

Professor Allman smiles: "Certainly, philosophers claim metaphysics as their sole field of study. Nevertheless, modern physics cannot do without what allegedly belongs to metaphysics. - Do you know what the goal of metaphysics is?"

Dr. Albert Maupertius, the philosopher among the participants, who is greying in honour, feels addressed: "It is about knowledge of the basic structure and principles of reality! - But that is actually my area of expertise, colleague, not yours."

"I may write down your words:

Knowledge of the basic structure and principles of reality!

Professor Allman presses a few keys on a computer keyboard and, with the help of a 3D beamer, projects the sentence glaringly brightly into the middle of the classroom. At the same time, the window panes become discoloured and darken like the lenses of sunglasses in order to dampen the sunlight. The second statement of his colleague, which is similar to a reprimand, he does not take into account.

"Moreover, the Greek philosopher Plato said that metaphysics is what comes after physics," adds Dr. Krates, the bearded assistant to Dr. Maupertius.

"That is also correct!" comments Professor Allman. "Classical metaphysics deals with questions such as 'why does anything exist at all' or 'what constitutes reality as such'. These are the questions we will also be dealing with. In this week of the course we want to deal with topics that are beyond the classical nature ofscience, but nevertheless have a meaning for scientific research.

The mathematician Dr. Anaximenes persistently objects: "Nevertheless, I would like to recall what the philosopher Kant said and what I am also familiar with as a mathematician. He believed that any attempt to formulate theories about reality that lie behind the things of experience is doomed to failure.

"Kant was right in his environment at the time," noted Professor Allman, looking first at Dr. Anaximenes firmly and then at the round. "Whether his statement is still valid today, I cannot at this point give a credible answer."

Faces that have been entered look back. If all objections to the course are to be correct in the first place, why are they sitting here?

"But," Professor Allman continues. "I don't think you came to this course You are keenly interested in the basic questions of your being. They want answers and not just any answers, but plausible answers. Answers that will satisfy you. Answers beyond religion, pseudoscience and esotericism. Answers on a scientific level. Short, credible answers based on the scientific method."

The grey-haired Dr. Maupertius replies: "That's exactly how it is, colleague! It is not philosophical expertise that interests me, because that is part of my profession, but credible answers to the fundamental questions of our being. I think I also speak for the rest of the audience when I say that the term metaphysics in the seminar announcement by a physicist confused us very much.

"Then let me briefly anticipate the outcome of my subsequent remarks: Without metaphysics there are no theories and no scientific method. In addition, I want to show you a way to find credible answers to the basic questions of being."

"That seems to me to be a contradiction, Professor Allman. For many people what their religious leaders tell them is already credible. Logic and empirical evidence do not count for these people, because they do not dare to think for themselves, but take over everything that is put before them. Other people are satisfied with esoteric descriptions. For them, what sounds comforting is already credible. For professional reasons alone, I belong neither to the first nor the second group of people. Precisely because I am a spiritualscientist, I want my critical mind to be satisfied. For this I need empirical evidence for the statements about the basic questions of being, i.e. something philosophy is not able to deliver. So I'm curious how you intend to solve the problem of credibility."

"I'm sure this course will satisfy you, Dr Maupertius. - Now to the topic: At the times when we humans still went hunting with bow and arrow, the knowledge of physics was limited to everyday life. The knowledge of why something worked originated from magical thinking at best. For everything there was some god or spirit that made the world go round. A theoretical basis was missing. Today, however, knowledge and understanding of the world is much more comprehensive. When we marvel at the breathtaking technical progress since the time of our fur-clad ancestors, and especially the progress of the last three centuries, we must remember that we owe it almost entirely to the 'scientific method', i.e. experiment, observation, logical thinking, hypothesising and refuting. What I have just summarized in a few terms belongs to what we call scientific theory."

Professor Allman slowly gets going and continues his introduction:

"I am well aware that you are all very well versed in the meaning of scientific theories. Nevertheless, let me discuss some basic questions on the basis of the following slide."

The glaringly bright beamer throws the film onto a smoky transparent curtain in the middle of the room. The curtain consists of a newly developed material. The image projected onto it makes the participants believe that it is floating freely in space.

> Seven criteria that must be met for it to be a <u>scientific</u> theory. It must be...
> 1. ... a reality can <u>be described</u> logically without contradiction, including the preconditions of this reality
> 2. ... this reality must be <u>explained</u> logically and, if necessary, further conclusions must be derived (=<u>hypotheses</u>)
> 3. ... an unnecessarily complicated explanation can be avoided, even if it is easier (<u>Ockham's razor</u>!)
> 4. The hypotheses must in principle be falsifiable (=check for falseness)
> 5. ... <u>empirically decide</u> whether the reality matches the hypotheses (falsify or verify)
> 6. Derivations of such <u>predictions</u> must be made which have practical significance
> 7. It must be possible to decide empirically whether the predictions are correct (falsify or verify)

Professor Allman looks around: "Can anyone give me an example of a scientific theory?"

"Yes, me!" said Johanna Balthasar, a strictly dressed woman in a grey dress and from the back row. The sign in front of your seat identifies you as a member of the theological faculty. "The world was created in six days by our Creator!"

A displeasing murmur goes through the auditorium.

"What makes you think this could be a scientific theory?" Professor Allman frowned.

"Because it is the absolute truth, after all, everything was written down literally as it comes from the Creator!"

"All right, let's check, using the seven criteria that make up a scientific theory, whether your statement is really, in fact, a scientific theory. Let's start with the first criterion. Is the young lady's statement the logically consistent description of a reality?"

Johanna Balthasar's right-hand neighbour, a tall young man whose name plate identifies him as Dr Benedict of Aniane, briefly takes a look at his neighbour and then replies: "In honour of the theological faculty, I would like to say that we too are committed to scientific thinking. As far as it concerns unprovable contents of faith, we do not want to lump our religion together with naturescience! As far as I know, the creation myth in Genesis 1,1 - 2,4a differs from the text in Genesis 2,4b - 25, which follows immediately after it: whereas in the first text the whole world is created first, and man does not follow until the sixth day, in the second text the individual acts of creation follow in a different order. Here the earth is dry at first, a barren steppe. Man is then created as an individual, then the plants and animals of the garden, so that man is not alone. So if one takes the text as a whole, I do not think that it is a logically consistent description of a reality. Already the first criterion of what constitutes a scientific theory is not fulfilled!"

"Not only that," exclaims Dr. Anaximenes, the mathematician from the front row, who apparently also has theological knowledge. "The first creation myth, according to which the universe is filled with water and the sky is a solid watershed, obviously contradicts demonstrable facts. After all, our spaceships do not float in water, but pass through the vacuum prevailing in space. Criterion five is not met. The hypothesis that the universe is filled with water does not fit reality."

"Very well analyzed," rejoiced Professor Allman. "If there is only the description of a reality that is perhaps not even logically consistent, then it is religion, pseudo-science or esotericism. This is not meant in a judgmental way, Dr. Benedikt von Aniane, but only as a classification. - But now I really want to hear from you the example of a scientific theory!"

The tall Dr. Aniane answers: "If six sheep graze in a pasture and seven are added, then there are thirteen sheep in the pasture.

His neighbor Johanna Balthasar blushes in annoyance: "This is supposed to be better than what I said before?

In the rest of the auditorium, restrained laughter can be heard.

With the words: "All serious contributions to the discussion are allowed! We don't want any prohibitions on thinking and no suppression of opinions", Professor Allman ends the laughter. "Both participants of the theological faculty are definitely serious contributions. - Dr. Aniane, will you please substantiate your statement?"

"Criterion 1: The thirteen sheep in the pasture are the description of a reality. One of the premises of this description of reality is that sheep exist and graze on the pasture and are not just imagination. Criterion 2: The hypothesis is that arithmetic can be used to add up the number of animals and thus arrive at a result. Criterion 3: There are no unnecessarily complicating explanations. For example, I don't need gods to reach the number thirteen. That means Occam's razor is done enough. Criterion 4: The hypothesis that simple arithmetic can be used to determine the number of sheep is verifiable. Criterion 5: I saw a sheep pasture behind the dunes, so we can decide empirically whether reality fits the theory. Let's test the theory for ourselves and go outside!" With those last words, Dr. Aniane stands up.

"Thank you, Dr. Aniane, please sit down. We don't have to go outside. Life experience speaks for itself," Professor Allman interjects. "But please proceed with your statement of reasons."

"Criterion Six: I can make predictions from the hypothesis. One of these predictions is: If there are thirteen sheep in the pasture and three come into the fold, then there must still be ten sheep in the pasture. Criterion 7: The predictions can also be empirically tested. As the shepherd in the pasture assured me before the start of the course, the predictions about the number of sheep have always come true!

Those present are briefly amazed, then they begin to knock enthusiastically on their benches.

"A physicist could not have expressed it better, Dr. Aniane!" praised Professor Allman as the knocking died down. "With this you have actually given the example of a physical theory and given a clean foundation. And you showed me more. In metaphysics one asks in the most general way what exists. The three great physicists Einstein, Podelski, Rosen, on the other hand, have given physics and thus also metaphysics a concrete criterion to decide when an element of physical reality exists. Sloppily speaking, they said that if the predictions have a probability of one, that is, if they are certain to come true, then there is an element of physical reality that corresponds to the predictions. To express the result metaphysically: Sheep exist! - We first had an example of a statement about reality that satisfies only criterion 'one' of a scientific theory, and now we have an example that covers all seven criteria and thus reaches the highest level of credibility. Is there anything in between?"

Dr. August Dessoir, a corpulent parapsychologist wearing nickel glasses, says: "I am thinking of scientific theories that have the potential to become accepted scientific theories, but where important things are still missing, such as the empirical decision of the hypotheses. The criteria 'one' to 'four' of scientific theories would then be fulfilled, but the criterion 'five' is lacking.

"Dr. Dessoir, are you thinking of your own field, parapsychology?" Professor Allman asks.

Dessoir adjusts his glasses: "I first studied geology before I became a parapsychologist. Therefore, I would like to take

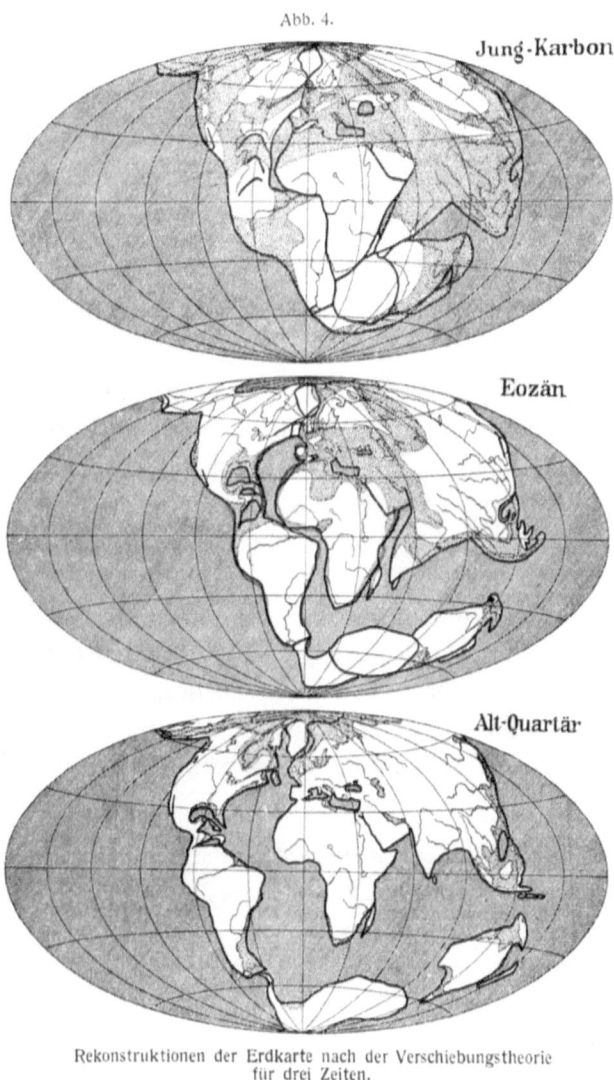

Rekonstruktionen der Erdkarte nach der Verschiebungstheorie für drei Zeiten.
Schraffiert: Tiefsee; punktiert: Flachsee; heutige Konturen und Flüsse nur zum Erkennen. Gradnetz willkürlich (das heutige von Afrika).

Figure 1: Wegener's displacement theory

Wegener's continental drifthypothesis as an example, which for a long time was regarded as pure speculation and finally, after its confirmation, was accepted by geology, i.e. by a recognized science".

"Thank you, Dr. Dessoir. Since not everyone knows Wegener's theories, I would like to start by playing a short film from the digital archive."

Professor Allman presses a few buttons and seconds later Alfred Wegener's projection image flickers up, seemingly floating freely in space. Professor Allman fast-forwarded the summary of Wegener's life, who died at the age of fifty in 1930 in Greenland during a research trip, to the point where Wegener's theory of continental drift is shown: A globe shows a single connected primordial continent like a huge island in water. After some time, cracks appear in the country and one can almost see the shapes of Africa, America, Australia and Antarctica. At the cracks the land is slowly separating. The continents split off from the primordial continent are drifting to their positions known today.

At this point Professor Allman stops the film. "Up to now we have seen the description of a reality, that is, what constitutes the criterion 'one' of a theory. - Let us now come to the explanatory statements! Dr. Dessoir, do you know what Wegener's hypotheses were?"

"As far as I know, Wegener could not plausibly explain the cause of continental drift, but he cited numerous phenomena which were inexplicable without continental drift and which, in my opinion, were quite sufficient to confirm his theory.

"And what were these unexplained phenomena?"

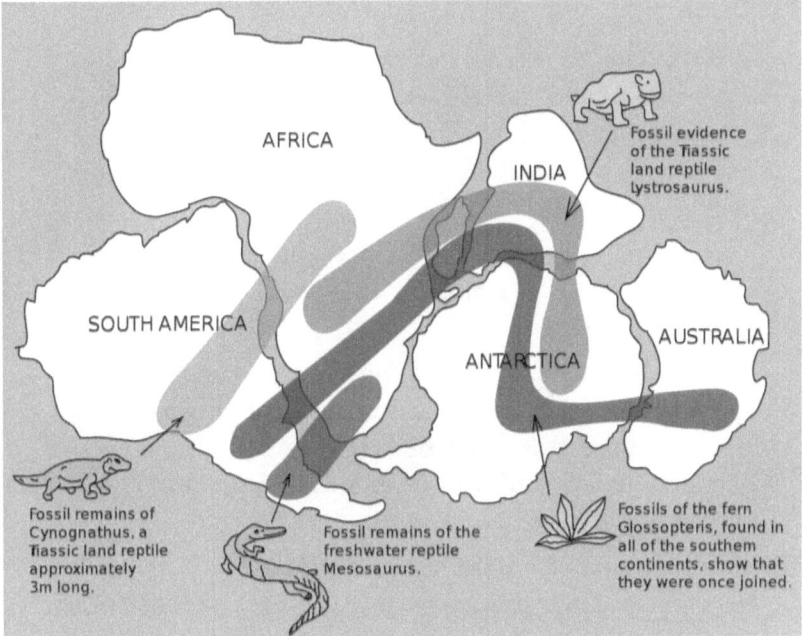

Figure 2: The palaeobiogeographic distribution areas of Cynognathus, Mesosaurus, Glossopteris and Lystrosaurus shown here allow the reconstruction of the original continent

"For example, the similarity of rock formations at the fractures of the continents in India, Madagascar and East Africa. There is a mountain range in South Africa, the extension of which can be found in a similarly constructed mountain range in Argentina, or Precambrian rocks in Scotland, which correspond to those in Labrador on the other side of the Atlantic. Norway and Scotland have folded mountains that continue into the Appalachian Mountains in North America."

"Well, these are phenomena for geologists. Are there any other arguments that Wegener's cited?"

"Of course, for example, those from paleontology. Fossils of a primitive reed fern (*Glossopteris*) and its associated flora have been found both in Africa and Brazil. Furthermore, the distribution of various reptiles in the different continents could be proven.

Professor Allman shows the palaeobiogeographic distribution areas of Cynognathus, Mesosaurus, *Glossopteris* and Lystrosaurus.

Dr. Dessoir continues: "And another important hypothesis concerned the climate in Antarctica. Coal deposits were discovered there that could only be formed under tropical conditions. In Wegener's view, all this could be explained by a continental drift alone."

"I understand: the amazing facts served to indirectly verify the continental drift hypothesis. Its value corresponds to that of a direct empirical decision of criterion 'five'."

"That's Professor Allman! If one were to deny the continental drift, all arguments would really be nothing but inexplicable phenomena. How else could one explain the coal deposits in Antarctica, for example? Unfortunately his contemporaries did not want to acknowledge this and so the continental drift theory fell into oblivion after his death."

"And what's the theory today, Dr. Dessoir?"

"No problem! Today, continental drift can be directly detected by satellite geodetic measurements. It is now even possible to predict by how many centimeters per year America will drift away from Europe."

"Then I may summarize: Wegener put forward a theory in which the continental drift could not be directly verified. That is, criterion 'five', the direct empirical decision as to whether his hypothesis is correct, was missing. This also meant that he lacked the credibility and recognition of most of his contemporaries. At best, his theory was regarded aspre-scientific. However, his contemporaries overlooked the fact that numerous otherwise inexplicable phenomena dependent on the hypothesis indirectly verified it. - In physics, too, we can only indirectly prove many hypotheses. This means, especially for the hypotheses of the metaphysical questions which we are dealing with here in the course from the point of view of physics, that they are indirectly verified.

"What do I need to hear?" Dr. Krates, the assistant to the philosopher Dr. Maupertius, jumps up from his seat and gets excited: "You lured us into this seminar under false pre-conditions, Professor Allman! - I thought it was a science event that had evidence to offer. Indirect evidence does not count in this context! Why should I believe in a physical reality that I can't see?"

"Can you see electricity, Dr. Krates?" Counters Professor Allman.

"Uh, no, not really", Dr. Krates replies sheepishly.

"And do you believe in the existence of electricity or not, Dr. Krates?"

Dr. Krates is sitting down again. "Well, I guess without electricity their projector up ahead wouldn't work either. But when it comes to things other than our everyday experience, namely the basic structure and principles of reality from a physical point of view, I expect direct proof before anything seems credible to me.

"Then let us take atomic nuclei as an example, Dr. Krates. Atomic nuclei are extremely tiny and have nothing to do with our everyday experience. Does the existence of atomic nuclei seem believable to you?"

Dr. Krates scratches his head in embarrassment. "I do think you physicists can show me atomic nuclei. In that respect, their existence seems to me to be plausible."

"Then I will inform you that the original proof of the structure of atoms could only be provided indirectly." Professor Allman looks around. "Do we have a particle physicist here who can give Dr. Krates an understandable description of Rutherford's scattering experiment of 1906?"

From the middle of the auditorium, Professor Ernest Geiger reports: "Before Rutherford, it was not known that the mass of the atom was mainly concentrated in a small nucleus. Rutherford wanted to know more about the appearance of atoms and so he had gold foil bombarded with the alpha particles of a radioactive element. The alpha particles did not all pass through the gold foil in a straight line, but were partially scattered, namely whenever they hit the atomic nuclei of the gold atoms.

While Professor Geiger explains, Professor Allman starts an animation. A walnut-like structure is projected into the room. Pea-like spheres, which are supposed to represent the alpha rays, fly out of a black box in a straight line. Most fly by the walnut. Some peas hit the walnut and bounce off it. The angle of rebound varies depending on where the walnut is hit by the pea.

"Your animation is wonderful," praises Professor Geiger. "If you could now cover the walnut, Professor Allman, we would have roughly the situation Rutherford was in when he wanted to find out about the atom. After all, he knew nothing about the inside of the atom."

Figure 3: Ernest Rutherford at the time of his spreading experiment

Professor makes the walnut disappear in the animation. All you see is the bombardment of peas and how peas bounce off something invisible.

"Without the walnut in the way, the sudden change in direction seems inexplicable. But if you start from the hypothesis that an invisible object stands in the way of the peas, you can determine the size and shape of the object they bounce off. How would you do that?"

"It's really not difficult!" says the red-haired mathematician Anaximenes. "All you have to do is pick up the peas with an umbrella. Depending on where the peas hit the screen, one can calculate back from which point of the invisible object they bounced. And each of these criteria represents one pixel. If you get just enough points this way, you can get an image of the walnut-like object."

Professor Allman nods and takes the floor again. "Do you all see that the bouncing of peas off something invisible, would remain an unexplainable phenomenon, if one did not start from the hypothesis that there is an object, in this case an atomic nucleus, at the point of bounce?"

A nod of agreement follows in the hall.

"And now Dr. Krates, Rutherford could not see the atomic nucleus. Would you consider proof of its existence as direct or indirect evidence?"

"I really didn't know that you physicists had to resort to such tricks to learn about the world. But your animation impressed me. I will also allow for the indirect verification of the hypothesis with the help of otherwise unexplainable phenomena as an empirical decision for the theory criterion number 'five'.

Professor Allman looks around: "Do we agree that the description of a reality begins to become credible for us scientists when at least the criteria 'one' to 'five' of a scientific theory are fulfilled? Even if the verification of the hypothesis is indirect?"

Dr. Dessoir, the parapsychologist does not want to agree with this yet: "What about my field of expertise? - I often note the existence of unexplainable phenomena. Take telekinesis, for example, which is acknowledged to exist. I describe the reality of a newly appeared telekinetic phenomenon, but I cannot add any explanatory statements, i.e. any hypotheses about reality. Is the description automatically implausible because criterion 'two' is missing?"

"No!" laughs Professor Allman. "The absence of criterion 'two' only shows that you do not yet have a theory at all, and thus there is no need to answer the question of whether the theory is credible. The description and credibility of the phenomena as such is not affected by this. But once you have a theory, you can use the otherwise inexplicable phenomena as indirect evidence for the correctness of your hypothesis, just as, for example, Rutherford proved the atomic nucleus or Wegener proved the continental drift.

"Thank you, Professor Allman! - Do unexplainable phenomena actually occur in physics?"

"Yes, in quantum mechanics, unexplained phenomena are the rule. Let us take as an example what is known as wave-particle duality. In Rutherford's scattering experiment, the phenomena can be well explained if one assumes that the gold foil was actually bombarded with particles, i.e. tiny elastic spheres. However, if you make another experiment with particles, you get doubts whether they are really particles. This other experiment is the double-slit experiment."

Professor Allman projects a schematic image (Figure 4) of the experiment into space.

"In the double-slit experiment, the particles are fired from a source Q at an obstacle with two closely spaced permeable slits. In the example the particles are electrons. But the same results are obtained with other particles. - How many accumulations in the distribution of electrons on the photo plate behind the obstacle would you expect with two permeable gaps, Dr. Maupertius?

"I think it's two because the electrons fly through two gaps."

"One actually expects two clearly separated clusters of electron distribution behind the columns. This is shown in the upper part of the picture (Figure 4). But something else is happening. The distribution of the electrons on the photo plate shows considerably more than two bright areas. Instead, there are numerous light and dark stripes. It is a distinct interference pattern that is reminiscent of wave crests and valleys. If you throw a stone into a bowl of water in bright sunshine, the striped shadows of the resulting waves look very similar to the stripes that form on the photo plate (Figure 5). In addition, take a look at the illustration in the lower part of Figure 4. The crests of the waves in the drawing on the far right should mean a light stripe, the valleys a dark one."

"Have no mistakes been made in the experimental design?" wonders Dr Maupertius.

"The experiment has been conducted repeatedly for decades all over the world. Always with the same result! - I would like to state that objects like electrons, atoms, atomic nuclei etc. behave like particles, waves or both at the same time, depending on the experimental arrangement. If we take a close look at the photo plate, we see that the interference or wave pattern is made up of many small dots, just like the print

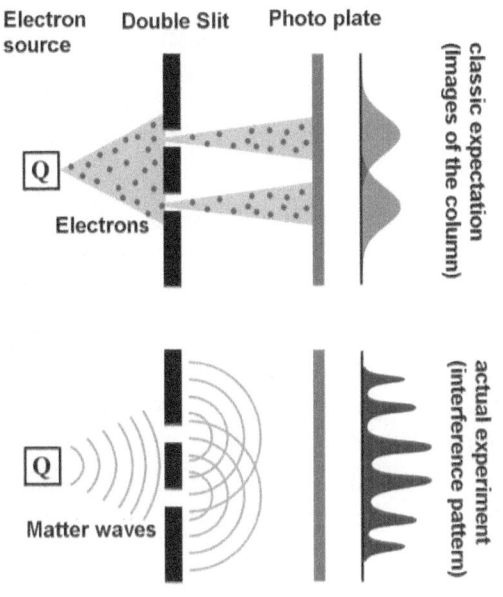

Figure 4: Double-slit experiment

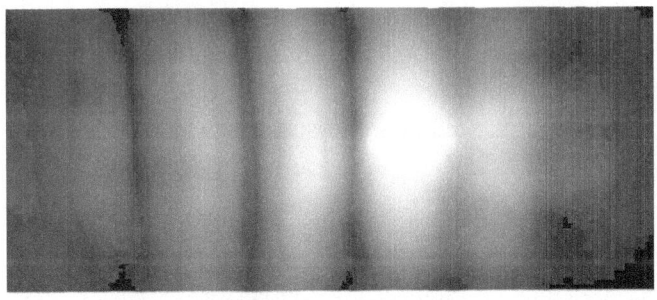

Figure 5: Interference bands on the photo plate

dots in an inkjet printer. We physicists speak of wave-particle duality."

"It seems to me it's really inexplicable. But you must have a theory on that, right?

"If you expect a clear and easily understandable theory, I must disappoint you at the moment. But if you accept a suitable mathematical model that belongs to the physical field of quantum mechanics, then there is something. Most quantum physicists consider the mathematical model to be a theory. This model performs excellently in the verification of the hypothesis, i.e. the criterion 'five'. One can derive predictions and decide them empirically, i.e. fulfil criterion 'six' and 'seven'. The hypothesis has never been falsified. Everything was always right."

"Professor Allman, can you briefly explain to non-physicists what you mean by quanta?" The theologian Johanna Balthasar is open to physical questions.

"With pleasure, Mrs. Balthazar! - The smallest known units in the physical universe are packets with a certain amount of energy. These packets we call quanta. Examples of quantum objects are the electron from the double-slit experiment earlier or the light particle photon. Quantums behave extremely strangely. We will talk about the peculiarities of the quanta in more detail in the course of this seminar."

Dr Maupertius frowned. "I heard a slight skepticism when you mentioned quantum mechanics, Professor Allman. What's wrong with a mathematical model if it works so well?"

"Let us first assume the double-slit experiment. As long as one has no statements about reality, it is an inexplicable phenomenon. If there is a hypothesis, the result of the double-slit experiment serves to verify the hypothesis. It is the indirect proof of their correctness. Criterion 'five' would then be fulfilled. - Now let us assume that a suitable mathematical model falls from the sky, but nothing else that is needed for a theory. Is it the description of a reality or is it the explanatory statement about reality, i.e.: hypothesis?"

Dr. Anaximenes, the mathematician feels addressed: "Apart from the fact that mathematical models do not fall from the sky, I think it explains the connections. The formulas also make it easy to make predictions. It must be the criterion 'two' of the theory, i.e. the hypothesis!"

"Well, Dr. Anaximenes! Who or what describes reality (= criterion 'one')?"

"You said nothing more fell from the sky, Professor Allman!"

"That's right!"

"If you have nothing else, I would use the mathematical model to describe reality!"

"Before answering this, I would like to remind you of the example with the sheep of Dr. Aniane. What was criterion 'one' there?"

"If six sheep are grazing in a pasture and seven are added, then there are thirteen sheep in the pasture afterwards, provided that sheep and pasture exist.

"Yes, and the criterion 'two'?"

"You can use arithmetic to add up the number of animals and get a result."

"Now take the same statement for criterion 'one' as for criterion 'two'. What falls under the table?"

"Heavens, the scales have fallen from my eyes, Professor Allman. If the original criterion 'one' is removed, then one no longer knows anything about sheep. The hypothesis that arithmetic can be used to add the number of animals is no longer a statement about an underlying reality. One still has a mathematical model, arithmetic, but a model should always be the reflection of a reality. I therefore ask myself, for which reality should arithmetic be a model, if there is no more descriptive statement about reality".

"As you can see, Dr. Anaximenes turns the original theory into something I no longer want to call a theory about the sheep in the pasture. It has become at best a mathematical predictive model for the addition of animals."

"So what does this mean for the theory of the double-slit experiment?"

"Quantum mechanics with its mathematical model remains an excellent tool for predicting the outcome of quantum mechanical experiments such as the double-slit experiment. But if one takes the catalogue of criteria of a scientific theory seriously, then quantum mechanics is scientific, but not a theory. Criterion 'one' is not fulfilled. It does not make any statements about an underlying reality but only predictions. In this sense, the outcome of the double-slit experiment remains an inexplicable phenomenon, just like in almost all other quantum mechanical experiments.

Dr. Maupertius interrupts: "You mentioned the term wave-particle dualism, Professor Allman. You physicists seem to know the nature of electrons, so you should have a description of the underlying reality.

"Even though physicists speak of wave-particle duality, the electrons sent through the double slit are in reality neither particles nor waves. We simply do not know what the essence is of what in the experiments once shows the property of particles, another time that of waves. We only know it exists."

"I am beginning to realize that physics needs metaphysics to describe what constitutes reality as such."

Professor Allman nods. "And this brings us back to the subject of this course. Physics and metaphysics are not enemies, but two married couples. Physics needs metaphysics when it comes to describing a reality, because this is often what physics lacks. Conversely, metaphysics needs physics if it is to prove the existence of what constitutes the essence of reality. In this event we will try to look at the results of all the individualsciences in a unified view and to create a metaphysical world view. The aim is to formulate a theory about our destiny as human beings. Not a theory that is commonly understood to be contrary to practice, but a scientific theory. Even if in the end we were to meet only criteria 'one' to 'five' and thus not fully meet the strict criteria of a scientific theory, we would at least have a credible description of a reality as a result, because it was verified by criterion 'five'.

The particle physicist, Professor Geiger, says: "I am practice-oriented, Professor Allman. Therefore, I would like to know how you intend to proceed in concrete terms? Will you float with us in philosophical heights or do you have something solid to offer us?"

"There is no shortage of solid evidence, Professor Geiger. We take as a basis, among other things, the many inexplicable phenomena from experimental physics, which you are no longer surprised about just because you come into contact with them every day. We look for appropriate metaphysical explanations of reality, build a theory from the ingredients and use the phenomena to verify the hypothesis. In the end we will have answered the basic questions of our existence."

Professor Allman looks at his watch. It's just after 12:00. "That late? - Let us take a short lunch break here and treat ourselves to a refreshment before we take a look at other topics at 14:00."

What comes after physics

> *God, freedom and immortality of soul are*
> *the tasks that*
> *all preparations of*
> *metaphysics, as their ultimate and only*
> *purpose, aim*
> *to solve.*

Immanuel Kant, Critique of the power of judgement

Monday 2 June afternoon

"Next, we will not discuss a theory, but first collect unexplained phenomena. Within the framework of these phenomena, questions will come up. Some of these questions we can answer right away, some later, some not at all. We will develop ideas about what might be behind the inexplicable phenomena. Only then do we try to bring everything together in a theory in order to get certainty as to the extent to which our ideas are credible. - If we are looking for the last level of explanations, if we want to get behind what makes up the reality of the cosmos as such, if we want to explore what comes after physics, then we have to look beyond the area that belongs to experientialscience. But how can one catch a glimpse of what is apparently not experiencable? For this we need a trick, which I would like to introduce with a parable."

Professor Allman is starting a movie. The start picture shows the Greek philosopher Plato. A speaker in the

Figure 6: Plato

background reads the original text. While he speaks, his statements are illustrated by moving scenes.

> *Plato: The cave allegory.*
> *Description of the situation of the prisoners*
> *(translated by Friedrich Schleiermacher).*
>
> *Next, I said, compare our nature in terms of education and ignorance to the following state. See*

people as in an underground, cave-like dwelling, which has an entrance open to the light along the entire cave. In the latter, they have been tied up at the neck and thighs since childhood, so that they remain in the same spot and only look forward, but are unable to turn their heads because of the bondage. But they have light from a fire that burns behind them from above and from far away. Between the fire and the prisoners there is a path, along which there is a wall, like the barriers that the jugglers erect in front of the spectators, over which they perform their tricks. - I see, he said. - See now along this wall, people carrying all kinds of utensils, and statues and other stone and wooden pictures, and all kinds of work; some, as if naturally, talk, others are silent. - A very strange picture, he said, you represent strange prisoners. - Very much like us, I replied. For first, do you think that such people have ever seen anything other than the shadows cast by the fire on the wall of the cave opposite them? - How should they, he said, when they are forced to keep their heads immobile all their lives! - And of the one before him, not this one? - What else? - Now if they could talk to each other, do you not think that they would also be in the habit of naming what they saw? - Necessary. - And how, if their dungeon also had an echo from over there, do you think, if one of the passers-by spoke, they would think they were talking something other than the shadow that just passed by? - No, by Zeus, he said. - In no way, then, can they take anything for real other than the shadows of those works of art? - That's impossible. -

At that moment, Professor Allman turns off the film and begins with questions: "How do the prisoners see their reality?"

Dr. Maupertius, the philosopher, knows the answer: "The prisoners hold their cave for the whole world. They only see the shadows of events in reality or they hear the echo. Yes, they consider these shadows and the echo to be the whole reality. They have no idea what's real outside the cave."

"And for what is the story a parable, Dr Maupertius?"

"I think it is a parable for our own situation. We think the world we see and experience is also reality. But this world is for us like the cave of prisoners. We do not realize that much of what penetrates our limited world is information coming from outside about the true reality.

"What if someone told the prisoners that they see only the shadows of reality and that many shadows do not come from their own cave but from the much larger world outside?

"I think the prisoners would not believe it at first. Maybe at some point they would start to think. And then you would try to find out which are their own shadows and which are from the outside! Little by little, they would learn about the world outside that is happening behind them. I should like to emphasise it: The condition is that someone must make the prisoners aware that such a world outside exists."

"Thank you, Dr Maupertius. You have highlighted the artifice necessary for us to learn about what lies outside or behind our own limited world! - We must be ready to acknowledge that this reality exists behind everything we know. And we must become aware that the signals from outside are crossing the border to us. Then we will learn to differentiate the observed phenomena according to those that take place inside and those that take place outside our limitation. The information that reaches us across the border will allow us to draw conclusions about the big picture."

Professor Geiger, the particle physicist, has tasted blood: "Tell me, Professor Allman, where do you think I'll find the limit?" he asks excitedly.

"Step lively, Professor Geiger! First we need two new terms! Tell me, Professor Geiger, as a physicist, how would you describe the field of science that is part of our experience?

"There is a term belonging to the theory of relativity called 'space-time'. However, one would have to say a few explanatory words about it first. I think the term 'space-time-universe' would be more appropriate at the moment! - Maybe RZU for short."

Professor Allman writes the new term on foil, paints an oval around it and projects the result in the middle of the room. Then he turns to Professor Geiger again: "If you now think of another term for the entire field, then I could complete the presentation.

"I propose the Continuum."

"And now we need a term for the area of the continuum that does not belong to the RZU."

"Vacuum" is the right word. However, I would like to emphasize that the term as we have defined it does not correspond to the physical concept of quantum vacuum, because the properties do not match.

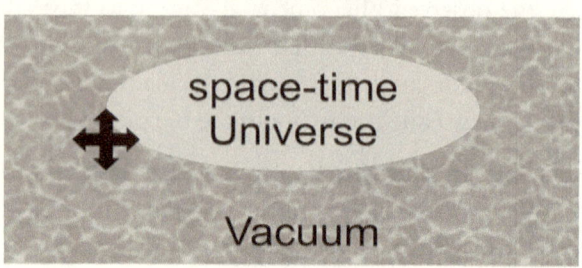

Figure 7: Continuum, vacuum and space-time universe (RZU)

"What properties we assign to the vacuum will be seen in the course of the course. I still like the term and so I will use it with pleasure, even though we may not be able to achieve complete agreement with the term quantum vacuum." Professor Allman completes his drawing.

Someone who has not yet come forward, the computer scientist Paul Aiken, asks a question: "Why is the space-time universeembedded in the continuum in your diagram, Professor Allman?

"I think we should define the 'Continuum' as 'the **totality of all that exists**'. That's why the RZU's scientific department, which is part of our experience, is apart of it."

"And what does the black crossed arrow mean?"

"The RZU is not isolated. Information penetrates from the vacuum into the RZU area. Information also goes in the opposite direction. The crossed arrow is meant to symbolize this flow of information."

"If I understand you correctly, we humans are located at the RZU and we should learn to distinguish how and where information enters or leaves the RZU from the vacuum.

"That's exactly how I imagine it!", Professor Allman is pleased to see Aiken's thinking.

Aiken remains skeptical. "But where exactly is this line?"

Professor Allman explains: "To answer your question, I must first discuss another double-slit experiment. This time I will explain the experiment with single light particles, i.e. photons instead of electrons. A metal plate with two narrow slits stands in the way of the photons. An observation screen is mounted behind it. The light source Q is a small laser, similar to a laser pointer used to explain presentations. The light from the laser can be regulated and attenuated to such an extent that only a single particle of light leaves the laser every second. The question of what happens when, for example, the full laser beam is directed at the double slit can be easily answered". Professor Allman projects the schematic image of the experiment (Figure 8) into the room. The middle and lower graphs remain covered for the time being.

"You can see it entered in the schematic above. The interference fringes occur as we know them from the double-gap experiment with the electrons. The wavy line to the right of the observation screen should represent the distribution of brightness. The crest of the wave means 'light strip' and the trough of the wave means 'dark strip'."

Aiken is curious: "What pattern is created when you cover one of the two columns?"

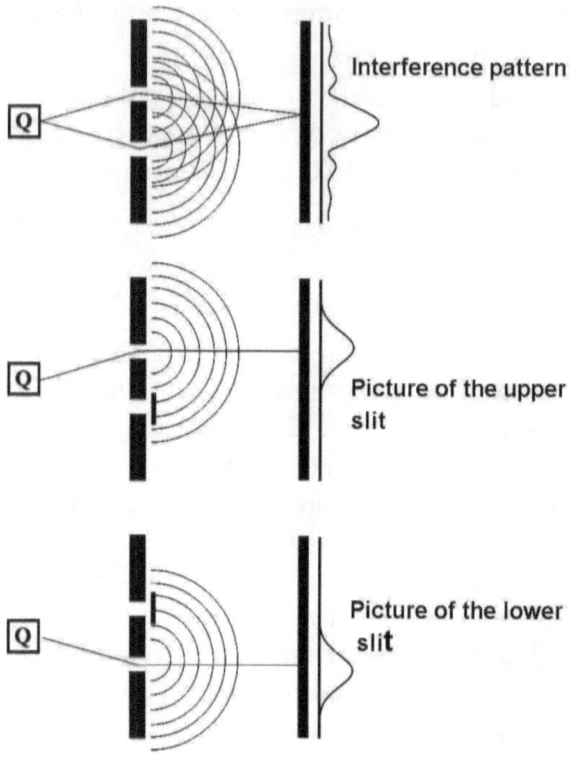

Figure 8: Double-slit experiment in different variants

"If I cover a gap, the interference bands disappear. There is only one bright stripe left." For explanation, Professor Allman shows the rest of the graph and then continues with his presentation.

"Now comes the interesting question: What happens if you attenuate the laser to such an extent that only one single light particle leaves the source per second? Will there be interference?" Professor Allman pauses and looks questioningly into the course room. The participants have to think for a moment.

The physicist Professor Geiger smiles because he knows the experiment. But he does not want to spoil the guessing of the other participants.

Finally, Dessoir, the parapsychologist. "I think there's nothing mysterious about the attempt. The interference, can certainly only occur when many photons influence each other. I imagine it to be similar to water. A single tiny drop of water cannot form a wave. In my opinion, wave formation is caused by mutual influence of neighbouring water droplets or particles. But where would a neighboring particle be found if there is only one single photon per second. After all, one photon cannot wait for the next to pass."

Dr. Anaximenes, the mathematician, argues astutely: "Another argument is that a single photon can only pass through a single slit. And what happens if the laser beam only passes through a slit, you have just shown us, Professor Allman. ...because then the interference stops."

Paul Aiken has other concerns: "I could imagine that individual photons shine so weakly that they cannot be seen with the naked eye. How can interference be detected at all under these circumstances?"

Professor Allman smiles. "Thank you for your committed and well-founded opinions and concerns. - I would like to start by answering Paul Aiken's last question: It is like a camera. If it is too dark, you have to expose long enough to still get a photo. This means: one only needs to wait sufficiently long, then in the course of time enough photons will hit the photo plate. The question whether the photons can interfere with each other is the subject of the experiment. But to keep you off the hook, I have a movie. This one I will play for you."

The audience is looking forward to the film, which is playing in front of them. First of all, they are shown the real test arrangement. The collecting screen for the photons is coupled with a computer. The initially black computer image shows in fast motion how one point of light after the other is added in rapid succession. Suddenly the room becomes restless. Someone says, "How is that possible?" Little by little, everyone recognizes it: the photons form an interference image with the stripes as known from the previous experiments. Professor Allman stops the film and looks around without asking a question. He's waiting for reactions.

"Is it really true what you have demonstrated to us, Professor Allman?" asks the theologian Dr. Aniane.

Professor Allman says one word: "Yes!"

Dr. Aniane goes on to consider: "Then how is it possible that the photons interfere with each other. When a slit is sealed, there is no interference image, but only a single bright strip. And a photon can only pass through a single slit, right?"

"We don't know whether a single photon passes through only one slit or through both at the same time," smiles Professor Allman.

"Isn't there any way to find out?"

"You can at least try to find out. That is the task of another experiment!"

Aiken shakes her head: "There is still more unresolved. Even if the photon should pass through both columns at the same time, which I think is impossible, it would have to find out beforehand whether both columns are open. Is it snooping around to find out?"

"Somehow the photon 'knows' whether both columns are open," Professor Allman replies succinctly. "But this knowledge it cannot have acquired within the space-time universe!"

Dr. Maupertius shakes his head in confusion: "What also puzzles me is this: If the photons pass through only one slit at a time, i.e. one goes through one slit, another through the second slit and so on, then the photons following with a time interval should know about their predecessors. In particular, you should know that they must not go to the dark places, but only where they can form the light stripes over time. - It's very mysterious!"

"The information about the predecessor and about the dark places where the photon is not allowed to go has not been obtained within the RZU. In a groundbreaking trial known as 'Bell's Inequalities', John S. Bell proved as early as 1969 that no theory of hidden connections within the RZU can do justice to quantum physics phenomena involving two photons," comments Professor Allman. "Therefore the information can only have reached the photon outside of what we experience as space and time. The only medium available outside the RZU is what we call vacuum."

"Is it even certain that light is made up of particles?" Dr. Anaximenes asks.

"This is certain, as many other experiments have shown for decades!"

"Then enlighten us, Professor Allman, what is the solution?" Dr. Anaximenes is getting energetic.

Professor Allman replies gently: "If the theory of an information exchange between two photons inside the RZU is not a solution, then the solution lies outside the RZU!

"You mean in a vacuum?"

"Yes, but before I work with you on a theory that explains the strange behavior of photons, I would like to offer you another experiment in which we try to find out which of the two columns the photons pass through."

"If it helps, I say clear the stage."

The film starts and initially shows the experimental setup again. In addition to the previous experiment, a small photon detector is now installed behind each of the two columns to find out which path the photons take. Each detector is coupled with an indicator LED. The experiment is running: As soon as the detector registers a photon, the LED lights up briefly. The audience is fascinated by what is happening. Only one of the LEDs is always lit. Never both light up at the same time. This means that the photons do not pass through two columns at the same time, but only through one column at a time. In fast motion a result is building up again. After it is clearly visible, Professor Allman stops the film. Stunned, those present stare at the finished picture. The interference is gone. There are only two photon clusters for the two columns (Figure 9).

"This can't be! Where have the numerous interference bands gone? Now that we know that each photon passes through only one of the two columns, it no longer behaves as before." Dr. Anaximenes is disappointed.

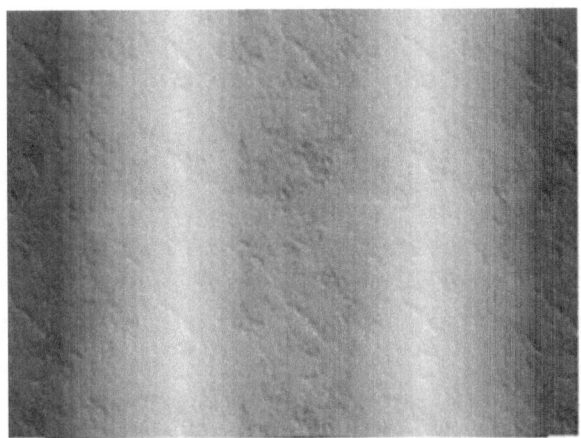
Figure 9: Two photon clusters without interference

"Why should what the photon does depend on whether we know which way it went?" Dr Maupertius shakes his head.

"First of all, I would like to point out that, metaphorically speaking, we have found the border we were looking for. We tried to find out something with a photon detector that the photon does not want to reveal. Here is a transition point from vacuum to the space-time universe. Photons, as we know, are quantum objects, and every measurement of a quantum means that it subsequently no longer behaves as it would have behaved unobserved and unmeasured. In the language of quantum physicists, the phenomenon of decoherence takes place with the measurement: a previously closed quantum system interacts with its environment. No one knows what properties quantum objects have before they are measured." Professor Allman takes a little pause for effect to let his words work. Then he continues: "Even if the quanta try not to give us any information about the vacuum, their behaviour allows certain conclusions to be drawn. Before we think about this, I would like to ask Professor Geiger to explain what else physicists have to say about the phenomena." With the words "Please, Professor Geiger," Professor Allman challenges his colleague.

"We physicists actually have numerous other experiments in store to find out how quanta and especially photons behave. For example, we have installed a shutter behind the double slit that moves up and down. At any one time only one gap is open. Also in this case no interference pattern is created. The same happens in any experiment, no matter how sophisticated, in which we could get information about the path the photon uses. It is not even necessary to retrieve the information about the route. The only decisive factor is that the information would, in principle, be available at the RZU. Whenever one conducts an interference experiment, i.e. when one does without the possibility of obtaining path information, the photon behaves like a wave, in the other case like a particle. This is what is known as wave-particle duality."

"Can we draw conclusions about the properties of a photon in a vacuum?" Professor Allman asks in between.

"We can see the behavior this way: In an interference experiment, the photon acquires the property of a wave, in a different experimental setup it acquires the property of a particle. We have to assume that the photon had none of these properties in vacuum before it was measured, **because both properties are mutually exclusive".** The last sentence was especially emphasized by Professor Geiger. "This means that the experimental setup adds certain properties or information to the photon. The transition from vacuum to RZU is located where this happens. This applies to all quantum objects."

With the words "thank you, Professor Geiger!" Professor Allman takes over the moderation again. "I'd like to record what you said about the vacuum." He writes the words on a new transparency:

<u>**Statements about the vacuum:**</u>

- No way-Information
- ...

Then Professor Allman turns to the auditorium: "Does this mean that in a vacuum there is no information at all?"

For a moment there is silence in the hall. After a minute the participants start whispering. Finally Professor Allman asks: "Please let us all participate in your discussions!"

Paul Aiken, the computer scientist, answers for everyone: "The photon must have existed before it was measured. So in a vacuum there is information about its existence before the transition to the RZU. In addition, we had found in the double-slit experiment that two photons do not inform themselves within the RZU. The information exchange takes place outside the RZU, namely via the vacuum."

"Good," says Professor Allman happily, before he goes on: "What do we call such a medium where the information is located until it is retrieved?"

"Do you mean an information store?" Aiken asks.

"That's who I'm talking about!"

"Then the vacuum would be an information store because the information about the existence of the photon and the information it needs for its subsequent behaviour is located there?"

"I think so!" comments Professor Allman and adds to his slide:

<u>Statements about the vacuum:</u>
- No way-Information
- Information memory
- ...

Professor Geiger's assistant, Dr. Robert Helmholtz, becomes restless.

"Is something Dr. Helmholtz?" Professor Allman wants to know.

"Yes, a path always occurs only in connection with a room. No wayinformation therefore means no spaceinformation. If this in turn means that the vacuum is a store of information outside space, then this store must also be outside time.

"Well, we have defined vacuum as that which lies outside the experiential space-time universe. So the vacuum is already by definition outside of space and time. The existence of an information store that lies outside of what we experience as space and time will seem so monstrous to you that I would like to confirm this view experimentally before we proceed with the elements of a theory of vacuum.

"And what is this experiment?"

"It is the double-slit experiment on the cosmic level! - Is there a cosmologist in this room?" Professor Allman looks searching.

Edward Michelson is like, "What do you expect from me, Professor Allman?"

"Could you explain how the cosmic version of the double-slit experiment with quasar 0957+516A,B works?"

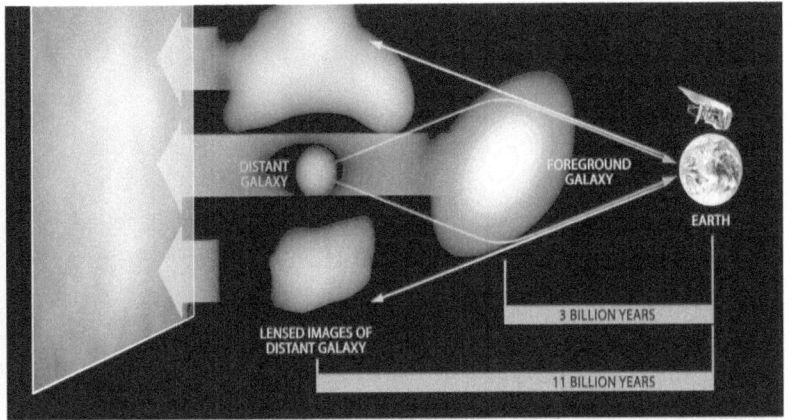

Figure 10: Gravitational lensing effect. At each of the apparent locations a quasar can be seen. The gravitational field consists of a galaxy cluster. The earth is the focal point.

"All right! - Quasars are often cosmic objects billions of light years away, which are extraordinarily luminous despite their great distance from Earth. It happens that there is a gravitational lens between the quasar and the Earth. This term is used to describe massive astronomical objects such as a group of galaxies whose gravity deflects the light of objects behind them. According to Einstein's general theory of relativity, the presence of mass bends space-time and thus also the path of light rays propagating in a gravitational field. Seen from Earth, it then looks as if a doublequasar is sending its light (apparent locations), when in reality there is only a single quasar.

Professor Allman projects the schematic image of a gravitational lens.

Edward Michelson explains further. "In the doublequasar 0957+516A,B, the photons on path one are travelling about 50,000 years longer than on path two. The question is whether photons of the two paths interfere with each other. What do you think, ladies and gentlemen?"

Dr. Krates, who sat there for a while with his eyes wide open, grumbled: "Why should photons still interfere when they are 50,000 years apart in time and have therefore travelled different distances? After all, photons seem to be very sensitive when they think they do not belong together!"

Dr Helmholtz smiles knowing: "I have to disappoint you, Dr Krates! The photons of the apparent doublequasar do not interfere with the time difference of 50,000 years. **They interfere with each other.** "He emphasizes the last sentence.

Astonishment fills the room. Dr. Krates is not satisfied with "Is it always like this?"

"Whenever the photons come from the same light source. However, if the doublequasar is real, i.e. the light comes from two different light sources, then there is no interference.

"How can photons, after billions of years traveling in different paths, know that they come from the same source?"

Professor Allman intervenes in the discussion: "The photons do not know this about any connection within the space-time universe, because there is no such connection. For this statement the already mentioned proof of John S. Bell, with his proof of 'Bell's inequalities', is valid again. - The photons can only **have received** the necessary information **outside of what we experience as space and time**. The cosmic variant of the double-slit experiment is thus a further argument for the property of the vacuum: "It is an information store outside the experiencable space-time universe.

An overwhelming silence spreads throughout the hall.

Finally the theologian Dr. Benedict of Aniane says something. "I can't shake the feeling that the vacuum has theological significance. But I still cannot get used to the fact that physics has found something existing outside of what we experience as space and time. It would help me, Professor Allman, if you had another impressive argument for such a vacuum!"

"Well, then I'm going to bring in the big guns. It is about what Albert Einstein called 'spooky long-distance effect'. This 'spooky long-distance effect' is the physical effect that enabled Professor Anton Zeilinger from Vienna to realize a science fiction dream, at least in its basic features. In 1997 he was the first to perform quantumteleportation with photons. Science fiction fans also refer to such a teleportation as beaming. - I will now explain to you step by step and without complicated mathematics an experiment on the spooky long-distance

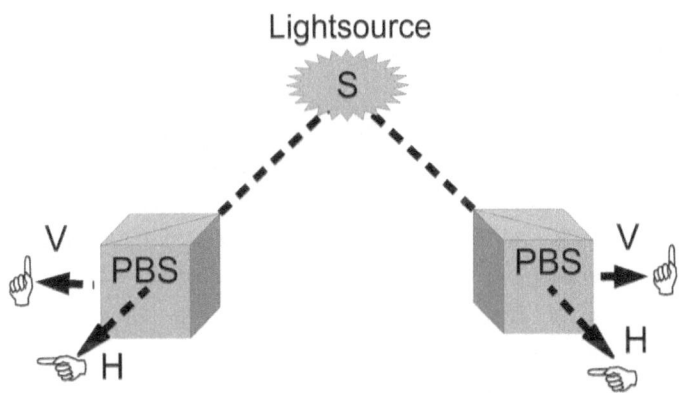

Figure 11: Two-photon experiment for 'spooky long-distance effect'; the light source S generates photon pairs with the property that their polarization is always rectangular; PBS=polarizing beam splitter

effect". Professor Allman projects a new graphic into the room (Figure 11).

He then asks his colleague Professor Geiger to briefly explain the term polarization. The latter is happy to agree to this.

"The experiment uses another property of light, polarization. This is a characteristic known from everyday life. Sunglasses often have a polarizing filter, for example to suppress sunlight reflections from the sea. The same applies to photographers. They use polarizing filters to block out certain reflections. Not only light rays but also single light particles carry polarization. If we imagine light as a wave, polarization is the direction of oscillation of this wave. This can point in vertical, horizontal or any other direction. However, the direction is always perpendicular to the direction of the light beam. One speaks of vertical or horizontal polarization or designates the polarization direction with an angle. To determine the polarization direction practically, the polarization filter is turned until the outgoing light is just as strong as the incoming light".

"Thank you, Professor Geiger. Professor Allman takes over the further explanation. "We come to the experimental setup. First there is a light source S, which emits photons in pairs. The photons show the property that their polarization is always perpendicular: For example, if one photon is polarized +45 degrees to the horizontal, the other is polarized -45 degrees, and so on. From +45 and -45 degrees results the right angle. Such a close connection of quanta as here the polarization of photon pairs is called **entanglement**. In practice, the entanglement is created with the help of certain atoms and a complex apparatus. However, we need not concern ourselves with the equipment any further. - Furthermore, we need two polarizing beam splitters (PBS) for the experiment. A PBS looks like a glass cube. The cube consists of two wedges glued together, each made of different crystal material. The polarization of light propagates at different speeds depending on the crystal material. This splits an incoming light beam into two beams that are polarized at right angles to each other. However, individual photons are no longer divisible, not even by a PBS beam splitter. Therefore, they have to decide which way they pass through the PBS, i.e. whether they come out vertically V or horizontally polarized H".

Professor Allman covers the right half of the graph. Only the light source S and the left area with the PBS beam splitter are still visible. "Of course, physicists have carried out many experiments in the past to find out how individual photons behave on a polarizing beam splitter. The result is that it is absolutely impossible to predict whether the photon will come out of the PBS horizontally or vertically polarized. Where the photon appears, whether at V or H, is purely coincidental. Because of the importance I want to repeat it: it is pure coincidence if a photon comes out of the polarizing beam splitter polarized vertically or horizontally. The same applies of course to the right polarizing beam splitter PBS. Now comes the completely inexplicable, which I would like to illustrate with an example."

Professor Allman projects an animation. There are two dice in a dice cup. The cup is shaken and poured out. The dice roll onto the table. They remain lying and you can see the number of eyes. One cube has two eyes, the other five. That makes seven eyes. The dice are rolled again. One cube shows three

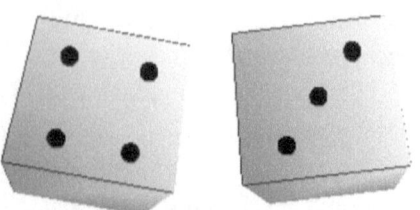

Figure 12: Interlocked dice always show seven eyes after the throw.

eyes, the other four, so again seven eyes. The dice are rolled again. Together again seven eyes. This is repeated many times. The dice always show seven eyes together.

"What would you say to such a dice result, dear colleagues?" asks Professor Allman.

"I think it can't be! This is completely unlikely," comments Dr. Anaximenes, who is well versed in probability theory. The other participants nod in agreement.

"Then watch the real experiment with the 'spooky remote effect' in the film now!" Professor Allman starts the projection.

The camera pans over the real test setup. A speaker starts to comment. Then a black line divides the film image into two halves. On the left you can see the left PBS, at each output H and V a photon detector, which flashes when a photon arrives. In the right movie picture the same thing only with the other PBS. Afterwards the experiment starts. The light source emits an entangled photon pair every second. One of the photons passes through the left path, the other through the right path. Left H lights up, right V. - Left V lights up, right H. Whether left V or H lights up remains random. But always the right photon shows the complementary polarization to the left one. Whenever H lights up on the left, V lights up on the right and when V lights up on the left, H lights up on the right. After a few minutes, Professor Allman stops the film.

"What do you say, ladies and gentlemen?"

"The two PBS will be so close together that the photons will be able to inform each other," believes Dr. Krates, assistant to the philosopher Dr. Maupertius.

"That's what physicists initially believed," answers Professor Allman. "Therefore, in the experiments, the distance between the left PBS and the right one was increased more and more. I know of an experiment that was carried out at a distance of 41 km, and yet the same result is obtained. The physicist aspect could prove with an additional sophisticated apparatus that the photons would have to spread the information about their path decision faster than light. But that's impossible."

The theologian Dr. Aniane leans back smiling: "I always knew it! There is simply no such thing as coincidence, everything is predetermined or with a technical term: everything is determined!

Professor Allman shakes his head: "John Bell was able to prove in an elaborate procedure that this is not so. Determinism cannot explain the phenomenon!"

The head of Dr. Anaximes is also working at full speed to find the solution: "There must be some hidden properties of the photon! Can it not be that the light source gives the photon pair a property that causes it to select the output from the PBS with the other polarization?

"Many quantum physicists have believed this for years and have developed corresponding theories. Then came the repeatedly mentioned John S. Bell and set up his 'Bell's

Inequalities'. Bell proved: "If we assume that no object in reality can act on another faster than the speed of light (term: locality) and if we maintain predetermination (= determinism), then really no theory worthy of the name can explain the results of thetwo-photon experiment.

Dr. Anaximenes is beside himself. Her eyes sparkle: "So how can it be that two random processes separated over a long distance produce a complementary result? - I wouldn't call it a coincidence."

"Please calm down Dr. Anaximenes!" Professor Allman appeases her. "In the experiencable space-time universethe 'spooky long-distance effect' of thetwo-photon experiment remains an inexplicable phenomenon, but somehow at least one of the photons of the experiment knows if and when its partner is measured and what result was achieved. Remember, we are like the prisoners in Plato's Parable of the Cave with one difference: someone told us that there is a reality outside our limited world. Therefore we find a solution. - It has been proven that photons cannot have experienced the information about the behaviour of their partners within what we currently understand by space and time. So they must have gotten the information outside the experiencable space-timeuniverse. Outside, according to definition, the only thing available is vacuum. **Therefore, the two-photon experiment or the 'spooky remote effect' is a proof of the existence of the vacuum.** "

Professor Allman silenced. The auditorium remains in reverent silence. The computer scientist Paul Aiken, who cultivates a modern vocabulary, speaks from the soul of everyone when he says just one word: "Wow!

A physical theory of the afterlife

*Beyond this world and this life groping
and searching is no longer possible.
There is only looking,
and everything looked at is truth.*

Joseph Joubert, thoughts, experiments and maxims

Tuesday, June 3rd

Professor Allman would like to give the course participants a breathing space after the research progress made the day before: "I think we should incorporate what we have achieved so far in a first and certainly still incomplete draft of a theory. Then we already have a working document that we can improve or extend. What do you think?"

Professor Geiger replies: "I agree with you. We should first write down the current state of affairs!"

Professor Allman also hears approval from the other participants.

"All right, let's do it this way! - Which of you would like to try to formulate the criterion 'one' of the theory? - When you begin as a philosopher, Dr Maupertius?"

Dr Maupertius nods. He reads his notes that he has made so far and then formulates them carefully. Professor Allman takes notes on his computer keyboard. The text is projected into the room for everyone to see.

Vacuumtheory:

Criterion one:

The part of the universe that is accessible to experiential science, which we want to call thespace-time universe (RZU), is not everything that exists. The reality that lies outside the RZU should be called a vacuum. This vacuum is different from the RZU.

There is a regular exchange of information between the vacuum and the RZU. Phenomena within the RZU, which are caused by the information exchange with the vacuum, allow conclusions to be drawn about the properties of the vacuum. Empirical decisions about its properties are made indirectly because of the definition of vacuum, because all experiencescience belongs to the RZU.

The reasoning is similar to that of Wegener in his continental drifttheory. One of the characteristics is that neither time nor huge distances play a role. The reactions of the vacuum are immediate and seemingly independent of what we experience as space and time".

"Thank you, Dr Maupertius, a physicist could not have done it better. On criterion two, however, I would have liked to hear Professor Geiger now.

Professor Geiger formulates.

Criterion two:

Hypothesis 1: The vacuum as we have defined it exists.
Hypothesis 2: There are phenomena at the RZU which allow a distinction to be made as to whether they are due to an exchange of information with the vacuum or can be explained completely within the RZU.

> Hypothesis 3: One of the properties of the vacuum is that of a store of information outside of what we experience as space and time. This memory can at least be tapped by quantum objects.

"May I also thank you, Professor Geiger. - What do you think Dr. Aniane. Are the hypotheses unnecessarily complicated, or is Ockham's razor simply done by finding the simplest possible hypotheses?

"I don't think it could be simpler. All other hypotheses, in particular those that start from a theory within the space-time universe, are probably not verifiable because of Bell's inequalities. As a theologian, I could introduce the hypothesis of a creator who is always sitting there, constantly busy using his omnipotence to produce the inexplicable phenomena of quantum physics. But I think such a hypothesis would only raise new questions and not really explain anything. It would therefore be the most complex answer imaginable and could not satisfy Ockham's demand for simplicity. Therefore, as a theologian, I am of the opinion that criterion 'three' is fulfilled!

Professor Allman noted:

> Criterion three:
> The hypotheses are the simplest possible. (Oral explanation by Dr. Aniane).

"By the way, in science up to now those theories have been crowned with success which found simple explanations for otherwise inexplicable phenomena. We are therefore on the right track! - Let us continue: what is the situation with criterion 'four'. Are the hypotheses verifiable in principle, i.e. can they be falsified?

Dr. Helmholtz laughs: "I have the feeling that you are asking a rhetorical question, Professor Allman, because we have actually already made empirical decisions. However, for the sake of form, I would like to reply seriously. Hypotheses one, two and three can be falsified by explaining all known phenomena within the space-time universe.

Professor Allman noted:

Criterion four:
Falsification possible if the known inexplicable phenomena within the space-time universe are explained. This means that the hypotheses are in principle testable.

"What can you contribute to criterion five?"

The cosmologist Michelson feels called to give the answer.

Criterion five:
Hypothesis 1 is indirectly verified by the two-photon experiment ('spooky long-range effect'). Without the existence of the vacuum, which serves as a store of information and which informs the quantum objects, it remains an inexplicable phenomenon how it can be that two random processes separated over long distances always provide the complementary result. All other explanations which remain within the framework of the experiential space-time universe cannot be verified.

Hypothesis 2: Before a quantum object is measured, it is obviously not an object of the experienceable space-time universe. The very attempt to measure a quantum object

> causes it to change its behaviour. This is shown by the double-slit experiments, which react sensitively to the attempt to obtain pathinformation. In this case the interference disappears. Therefore the measurement or the place where the measurement experiment takes place is a boundary between thespace-time universe and the vacuum. This verifies the hypothesis.
>
> Hypothesis 3: Without a vacuum, which serves as an information store, it remains an inexplicable phenomenon where the necessary information that explains the behaviour of the photons comes from. Therefore the hypothesis is indirectly verified.

"Excellent!" says a delighted Professor Allman. Then he makes a solemn face. "Ladies, gentlemen, colleagues. We have now reached the point where a theory credibly confirms what many of you have already felt: There is an afterlife. We physicists call this afterlife a vacuum. One more thing: Even though we have done nothing else than to make the previously unexplainable phenomena explainable in principle within the framework of a new theory, we have thus succeeded in making the hereafter accessible for empirical investigations. - But let us not rest on our laurels. To complete the theory, we need answers to criteria 'six' and 'seven'. Who wants to contribute?"

The parapsychologist Dr. Dessoir answers: "I have a question, Professor Allman. In my department there are more unexplained phenomena than explanations. Some things are very well documented. I have the feeling that these phenomena can be explained at some point with the vacuumtheory."

"I'm glad to hear that, Dr. Dessoir. Can you formulate a particularly impressive phenomenon so that it fits criterion 'six'? That means you would have to derive predictions based on your topic from vacuumtheory!"

"All right, I'll try!"

Professor Allman notes Dessoir's prediction in the computer.

> Criterion six:
>
> In the future it will be possible to prove that there is a transpersonal consciousness whose origin lies in the vacuum.
>
> The prediction is based on a series of controlled experiments in the field of thought and image transmission, which were already carried out in the early 1970s by the two physicists Russel Targ and Harold Puthoff.

Dessoir interrupts: "Professor Allman, I brought a film DVD with the experiment at that time! - Here it is!" He lifts the DVD up to show it. Then he gets up and brings her to the front.

"This is excellent!" rejoices Professor Allman and receives the DVD. He inserts it into the appropriate computer drive and immediately afterwards all participants can follow the experiment in the projection.

In the first scenes, the two physicists are introduced and the persons with whom the experiment is to be carried out. Then the scene changes. The person designated as the "receiver" now sits in a soundproof and opaque chamber that has been electrically shielded by the physicists. To record its brain waves, the receiver is fitted with a hood with electrodes. The derivation cables lead to an electro-encephalogram (EEG), which records the patterns of brain waves on a paper strip. The person referred to as the "sender" is led into another room. There she also receives a hood with electrodes attached to record her EEG. The transmitter is then exposed to bright flashes of light at regular intervals. Each time it flashes, the EEG records the special rhythmic brain waves that are usually caused by the exposure to bright flashes of light. The subsequent evaluation of the EEGpatterns of transmitter and receiver reveals a surprise. After a short time, the receiver's EEG shows the same patterns as the transmitter's, although it was not exposed to the flashes and could not receive any other direct signals from the transmitter. The speaker ends the film with the words: "How can we explain the transpersonal transmission of brain waves?"

Dr. Dessoir is curious: "What do you think, Professor Allman, will we be able to make an empirical decision about the prediction, criterion six, and tick off criterion 'seven'?

"Well, I think we need to know beforehand what the vacuum has to do with consciousness. This is definitely one of the topics of this course, but we will discuss other questions first. Therefore, the ticking off of criterion 'seven' must remain open for the time being."

"All right! Then I am curious about the connection between vacuum and consciousness."

Outside space and time

*In my opinion,
the cosmos is in us,
just as the other way round we are in the cosmos.
We belong to the universe as much
as he is a part of us.*

Yehudi Menuhin, art as hope for humanity

Tuesday, 3 June afternoon

Paul Aiken looks critically at Professor Allman. "On the one hand, the theory of the vacuum seems to me to be credible. On the other hand, your formulation 'outside of what we experience as space and time' sounds vague. Furthermore, I know that physics and mathematics are closely connected. Many predictions of physical theories are based on mathematical models. What about a mathematical model that specifies the vague formulation 'outside of space and time'?

"And I really can't imagine anything under it," criticizes the theologian Johanna Balthasar. "What the afterlife means, I know that, but your physical formulations don't fit!"

Professor Allman does not feel personally offended by the criticism: "Thank you for expressing your concerns! - It is true: the formulation I use is really spongy. I first wanted to awaken a basic understanding of the vacuum. After the first draft of the theory, it is time to make clarifications, improvements and extensions. It's just in time for you to point out a weakness." He turns searchingly to the rest of the auditorium: "Does anyone have an idea of how to be more precise on the one hand and how to illustrate the phrase 'outside space and time' on the other? His gaze remains fixed on the red-haired mathematician.

Dr. Anaximenes hesitates: "I already have an idea, but I would have to make a drawing before I can explain my idea!

"Why so timid? - Come up to the slide projector and just start drawing!"

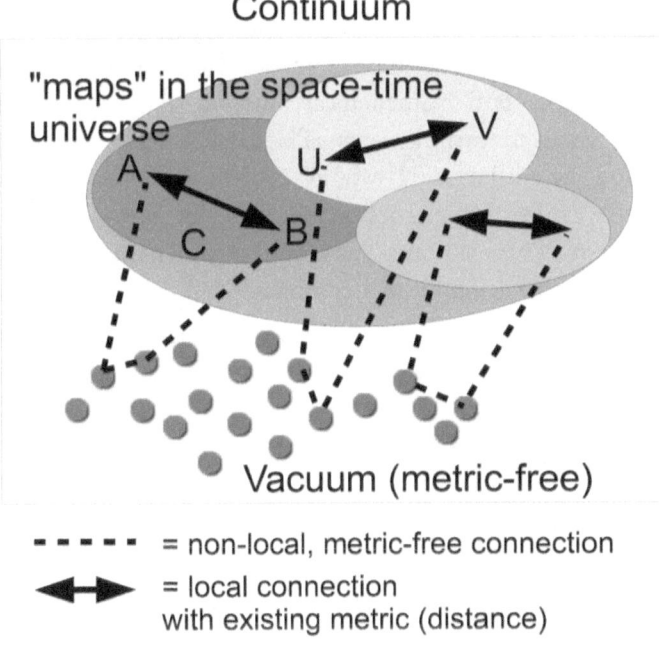

Figure 13: Mathematical model for the continuum (=continuum model)

Dr. Anaximenes will not be asked a second time. She comes forward and starts to draw an ellipse on foil within a large rectangle. This looks similar to Professor Allman's diagram of the continuum. As she draws, she mumbles incomprehensible words. Nevertheless, everyone looks excited and tries to gain a first understanding of what the diagram could mean. Finally her work is finished (Figure 13).

Dr. Anaximenes is beginning to explain "What I have just drawn is intended to represent a topological space. Such a space is a set whose elements can be understood geometrically as points and for which a few simple rules apply, which I do not want to discuss here further. - Some local subspaces have a metric. A **metric is** a mathematical function that assigns a numerical value to two elements of a room, which can be understood as 'distance'. I would like to call the local subspaces with existing metrics 'maps'. They are in thespace-time universe. There are also subspaces in which no metric exists. I think the vacuum is such a subspace."

Johanna Balthasar makes a sincere effort to understand what she has heard. "Can you give us a vivid example of what you're saying?"

Dr. Anaximenes thinks for a moment. "If you don't strain the example too much, I have one. Take a collection of road maps of all of Europe. Because Europe lies on the curved globe, it can only be shown in pieces on flat road maps. Within each map, the linear distance between two locations can be measured using a ruler and the distance calculated. It is not possible to measure the linear distance between two locations on different maps, because there is no metric for this. Even the general map, which shows the whole of Europe in a distorted form, does not allow a correct measurement of the linear distance between two places. If you try it nevertheless, the measurement leads to completely wrong values, because there is no metric here either. - Is it so far-fetched?"

"Thank you, I think I understand!" confirms Johanna Belthazor.

"In the diagram I have drawn connecting lines and used the term 'non-local' for the dashed lines. This means that the effect of one object on another cannot be explained by any effect that occurs at the maximum speed of light. Between points of the vacuum and the maps we have so far only discovered non-local connections in physical experiments. In this context I would like to remind you of the 'spooky remote effect'. As has been proven, photons obtain the information they need for their behaviour through a nonlocal effect. Distances seem to play no role at all in quantum experiments. From this I conclude that there is no metric in a vacuum. - For local connections within the maps of the space-time universe we have a metric. In the graphic this circumstance is symbolized by the double-sided arrow."

Professor Allman interrupts: "And how do you want to specify the vague formulation 'outside space and time', Dr. Anaximenes?

The red-haired mathematician smiles: "Oh, that's easy now: Outside of space and time is **the subspace of the continuum where no metrics exist.** It's the vacuum! This is nothing surprising, but it is precisely formulated."

Professor Allman thanks the mathematician: "You have developed a vivid model of our current knowledge of the continuum. In particular, the introduction of the term 'metrics' seems to me to be helpful for a better understanding." Dr. Anaximenes returns to her seat and Professor Allman continues.

"I would like to immediately record the finding in the list of statements about the vacuum." He's writing a new slide.

<u>Properties of the vacuum</u>
- topological **space without metrics** (distances do not matter)
- the space points are **information memories**
- ...

The philosopher Dr. Krates feels uncomfortable: "Professor Allman, a room in which there are no distances is for me something that contains nothing at all. In addition, the term vacuum associates with me an empty space. Is it possible that the vacuum we define is such an empty space?"

Professor Allman strokes his beard before answering: "We know that vacuum is an information store and there is more than just empty space behind an information store. What exactly is behind this, I actually wanted to investigate with you later. But it seems to me to be sensible to transfer part of this investigation already to this place. To do that, we need to clarify three questions." He notes on foil and projects:

1. *What theory does mainstreamscience have about the quantum vacuum?*
2. *What is the relationship between the quantum vacuum and our vacuum?*
3. *What is the essence of matter?*

"Let us begin by answering the first question." He looks around and sees his colleague Geiger getting restless. "I believe Professor Geiger, you'd like to take over."

Professor Geiger starts joyfully: "Einstein's theory of relativitysees only the geometry of space-time in empty space. In reality, it shows that the cosmic vacuum is by no means just empty space. Based on mathematical models of quantum mechanics, physicists have predicted that even at absolute temperature zero and in the complete absence of fields carrying matter or energy, empty space still contains energy. This means that something must exist in empty space, namely what we call vacuum energy. Although we cannot directly detect the vacuum energy, we are finding more and more interactions between this vacuum and the observable objects and processes of the physical world. Thus the vacuum energy is indirectly proven. Calculating the magnitude of this energy has caused us enormous difficulties in the past. Today, theory assumes an energy density of about 100 trillionths of a watt second per cubic millimetre of space volume. The associated theories state that the energy density is not proportional to the volume of space, but is highly proportional to the surface area of the volume".

"Thank you, Professor Geiger! - Dear colleagues, I would like to draw your attention to some of Professor Geiger's formulations. He said: "The vacuum energy cannot be proven directly! - What else did he say?"

Dr. Anaximenes reports: "He talked about interactions between the vacuum and the observable objects of the physical world."

Professor Allman nods contentedly: "That is well noticed. Then you will be able to answer me in which area the vacuum energy belongs. Does it belong to the quantum vacuum of the physical world or does it belong to the vacuum outside the observable space-time universe as we defined it?

"Clearly the vacuum energy belongs to the vacuum as we define it here, because it is not directly observable!"

"Thank you, Dr. Anaximenes! - I would like to make a note of this surprising result right now." Professor Allman projects the added slide.

<u>Properties of the vacuum</u>
- topological **space without metrics** (distances do not matter)
- the points in space are **information memories...**
- ... and contain the **vacuum energy**

Dr. Helmholtz does not agree: "But we have determined an energy density per volume of space for the observable physical world. While the vacuum as defined here has no metric. That means it has no spatial volume and no energy density."

Someone who has not yet contacted Professor Allman comes to his aid with arguments.

"My name is Enrico Fechner. I'm working on a vacuum energy project. I think the difficulties physicists have had in determining energy density so far are simply due to the lack of a vacuum model. Although we measure the interactions in the observable physical world, we must draw conclusions about the vacuum that cannot be directly observed. Think of Plato's Allegory of the Cave! **The shadows on the cave wall are not reality, but they tell everything about the real world.**"

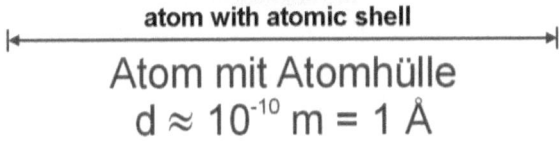

Figure 14: Atom model (not to scale)

Helmholtz is opening its eyes. "I had forgotten Plato again. You're right! The shadows are not reality."

"It's great how we can reach new insights together," says Professor Allman. "The second question has been answered. The third question remains, namely what is the essence of matter. First of all, I would like to show you a model of an atom. Who can explain it?"

Fechner says: "I haven't said much yet, let me explain!"

"I'd love to.

"It is common knowledge that matter is made up of atoms. What is less known are the constituents of an atom. It consists of an atomic nucleus and the electron shell, in which electrons circulate. Electrons can be everywhere with different probabilities, therefore the electron shell is not a sharply defined area, but is drawn as a nebula. The nucleus itself is about 10,000 times smaller than the electron shell and consists of protons and neutrons".

Professor Allman asks an interposed question: "And what are electrons, protons and neutrons made of?"

"Electrons no longer consist of anything else; they are among the smallest building blocks of matter, the so-called elementary particles. Protons and neutrons, on the other hand, are composed of two other elementary particles, the upquark and the downquark. Apart from the glue that holds the whole thing together, matter really only consists of three different components: "Electron, upquark and downquark.

Johanna Belthazor cannot believe what she hears. "But there are different atoms! I think of the gold atom and then I also heard of the hydrogen atom! The atoms are so different, there must be something else at play!"

"Not really! Both gold atoms and hydrogen atoms consist of the three elementary particles mentioned above and nothing else. The only difference is the number and arrangement of the elementary particles."

"I don't understand that!"

"Let's take an example. Imagine two children, a girl and a boy, have a bucket full of Lego bricks. The bucket contains three different types. Like stones with two, four and eight pimples. From these stones the girl builds a small doll's house with two rooms, furniture, oven etc. The boy, on the other hand, builds a large castle with mighty walls, battlements, gate opening and moat. What is the difference between a dollhouse and a castle?"

"You are right. The two structures do not differ in the type of stones, but only in the number and arrangement of the stones."

"The same is true for the different atoms! They differ only in the number and arrangement of the elementary particles!"

Dr. Krates argues: "But the statement really cannot apply to the wooden chairs here and the hamburger that I sometimes gobble down at lunchtime from hunger! - The two objects are too different. There must be something else that sets them apart!"

Fechner amuses himself: "I understand what you mean. You meant to say, "You'd never bite a wooden chair if you were that hungry."

The auditorium must laugh.

"But seriously, what atoms are the main component of wood?"

"I think it's the carbon with the chemical symbol C."

"So it is! - And which atoms are the main constituent of flour from which the hamburger buns are baked?"

"Um, those are carbon atoms, too."

"So it is! - I could ask the same question for each ingredient. In the end, we would always end up with atoms that consist of the three elementary building blocks. The atoms form molecules, which in turn differ from other molecules only in the number and arrangement of the atoms. The molecules form substances that differ from other substances only in the number and arrangement of the molecules. This continues on to the top level, the finished objects. Again and again, two different objects differ only in the number and arrangement of the building blocks. Ultimately, all the matter in the universe consists of only three different elementary building blocks.

Fechner finishes his presentation and the participants have to process what they have heard.

Professor Allman finally thanks the participants for their informative contribution and then asks a question: "We heard the term 'arrangement' over and over again. For the purposes of the course we need another term that means the same thing. So I ask you, what term can we use instead of 'arrangement'?"

The computer scientist Paul Aiken beams: "Surely you don't mean the term 'information'?"

"That's him, Mr. Aiken! - If, at the lowest level, all matter is made up of the same three elementary building blocks, and if one disregards the number of building blocks used, it is the form and meaning of this form that distinguishes two material objects. Remember the example of the doll's house and castle. Both differ in the form and meaning of their individual shapes. The one form means 'doll house' the other 'castle'.Information' is the general term for elements with a specific form and meaning. So we want to use the term 'information' instead of 'arrangement'. The three elementary components are the same everywhere and are practically interchangeable. Information is what distinguishes different objects. Therefore I would like to record the following on the projection foil:

INFORMATION IS THE MOST IMPORTANT COMPONENT OF THE SPACE-TIME UNIVERSE.

Dr. Maupertius replies: "For the sake of the other scientists who are not computer scientists, could you please define the term 'information' more precisely, Professor Allman?

"I'd love to. Professor Allman describes a projection foil with text and graphics:

> Information is defined as the pattern of a type of energy or the form of matter to which a certain meaning is assigned within a reality.

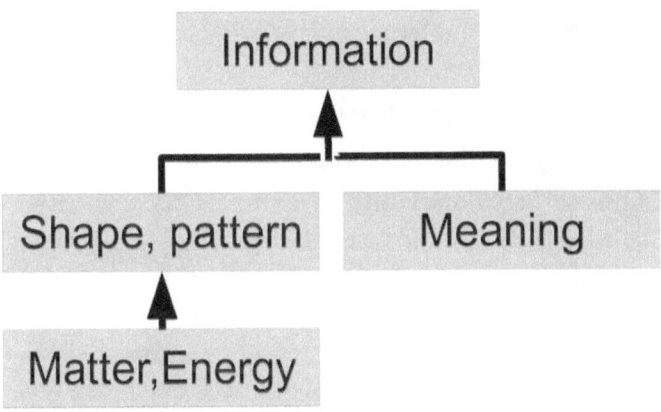

Figure 15: Components of information

"Does this mean that information and matter or energy are inseparably linked, because information is the pattern of a type of energy?" Dr. Maupertius asks for understanding.

"Yes, and as a result: A store of information is always matter or a type of energy."

Dr Maupertius protests: "To the best of our knowledge, the vacuum contains no matter!"

"Right! But we know: the vacuum is a store of information. From this it follows inevitably: it contains energy. Even if we had not known about the quantum vacuum of mainstream physics, we would have inevitably encountered vacuum energy."

Dr. Helmholtz strokes his forehead, probably to stimulate his thinking: "To return to the essence of matter and, taking into account Einstein's famous formula

$$E = m * c^2$$

which describes the equivalence of energy E and mass m: Can you say that

$$\text{Matter equivalent to information}$$

is because energy and information are inseparable?"

"I think so!" confirms Professor Allman. "One might even go so far as to conclude The vacuum is the source of all matter or of everything we know. But I do not want to pursue this approach for the time being. Back to the concept of information! It has a syntactic and a semantic level. Information consists of two components, each one of the two levels. First of all from patterns or forms, that is the syntactic level and secondly from their meaning, that is the semantic level! How about a little exercise, Dr Helmholtz?"

"With joy!"

Professor Allman holds up a piece of paper on which the following letters are written with a felt-tip pen:

TAU

"First of all I may note that this piece of matter has a certain shape or pattern. Please tell me how you determine the meaning of the pattern, step by step! What can it mean?"

"First of all, nothing! I need a frame of reference that the pattern is related to!"

"Through a reference system you already assign a fundamental meaning to the pattern. If you take the German language as a reference system, then the meaning of a word is already assigned to the pattern! - Alright, take the German language as a reference system!"

"In the German language, the letter pattern TAU can mean either precipitation, rope or Greek letter! To decide what it actually means, I would have to know more about the reality with which the letter pattern is related."

"So you need a reality! - However, this will not be enough to make a decision. Who or what do you need besides reality?"

"I think there must be someone there who observes reality and judges whether TAU means precipitation, rope or Greek letter. On the basis of the observation, a judgement must be made and thus the meaning assigned to the design."

"What is the process that makes the decision and assigns meaning to the pattern, Dr. Helmholtz? For example, can it be an automatic process like a computer program?"

Dr. Helmholtz ponders for a while before answering: "It must be a conscious process, because the automatic assignment of a meaning based on rules, as in a computer program, is not very accurate and repeatedly leads to stupid errors. A computer program simply does not recognize the real meaning."

"Thank you! That's all I wanted to hear! I get to hang on to it: For an accurate determination of the meaning, a **conscious process is** required. The latter assesses the pattern in the context of reality and assigns it a certain characteristic or representation of reality. The representation of reality then represents what we understand by meaning. - Information is closely linked to conscious processes or consciousness via the level of meaning. Without consciousness there is no real meaning and without meaning there is no information. **Consciousness is therefore the primary or comprehensive, meaning and information a part of it.** Professor Allman projects the latest findings in graphic form.

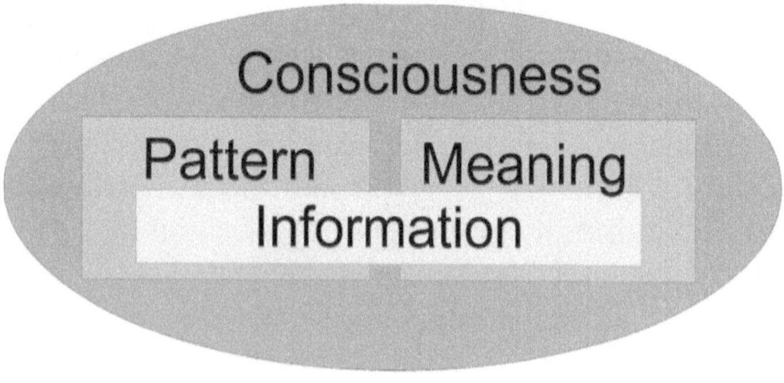

Figure 16: Information as part of consciousness

He looks at his watch. "Ladies and gentlemen! That's it for today! In our next topic tomorrow we will take a closer look at consciousness and the question of what the vacuum in its capacity as an information store has to do with consciousness. I wish you a pleasant evening."

The primary consciousness of the vacuum

> *In relation to consciousness,*
> *we do not speak*
> *of spatial indivisibility*
> *(since consciousness is not material and*
> *therefore not spatial),*
> *but of temporal*
> *indivisibility.*

Dalai Lama XIV, The Talks in Bodhgaya

Wednesday, June 4th

"Who is a consciousness researcher and would like to introduce us to consciousness?" Professor Allman looks around searching. His gaze remains fixed on a middle-aged woman with luxuriant frizzy hair. It is Professor Elisabeth Delacroix, who has already published several books on the subject of consciousness.

Professor Delacroix makes a defensive face: "The subject is very complex and there is no generally accepted precise definition of consciousness. But you need something precise in this course, if you do not want to end the treatment of the topic in general gibberish! Am I not right?"

"In your subject, different types of consciousness are distinguished, one of which is relatively easily accessible for the formation of theoretical models. There are also criteria by which you can recognize conscious behavior. Please, Professor

Delacroix, I would be very pleased if you could find a few words of encouragement."

Elisabeth Delacroix smiles and a few curly hair strands fall forward as she lowers her head. Immediately she looks up: "If you project images fitting to my speech, I agree!"

Professor Allman is looking for a symbolic start picture and finds one from the 17th century.

"Cogito ergo sum - I think, therefore I am! This was the famous saying and basis of the metaphysics of René Descartes, the 17th century naturalscientist whointroduced self-consciousness as a philosophical subject. Descartes did not yet make a distinction between thinking and consciousness. To him it was one and the same. In his time it was assumed that body and mind are two separate phenomena, in which only the body is accessible to naturalscientific investigation. Today we reject this. Mental phenomena are merely a special manifestation of natural processes and the mind is naturallyaccessible to scientific methods. For example, we can use magnetic resonance imaging to watch the brain think. We no longer need to find out how two substances as different as spirit and matter can interact. It is sufficient to find out which conscious or unconscious phenomena are related to biological processes. - Experts distinguish between two types of consciousness, the cognitive and the phenomenal. Cognitive awareness is an awareness of 'something', for example of the environment or of one's own body states. When the noise of the street penetrates my ear or I feel cold, it is a primary form of cognitive awareness. It consists of a representation of the environment and the own body."

"May I interrupt for a moment?" asks Professor Allman.

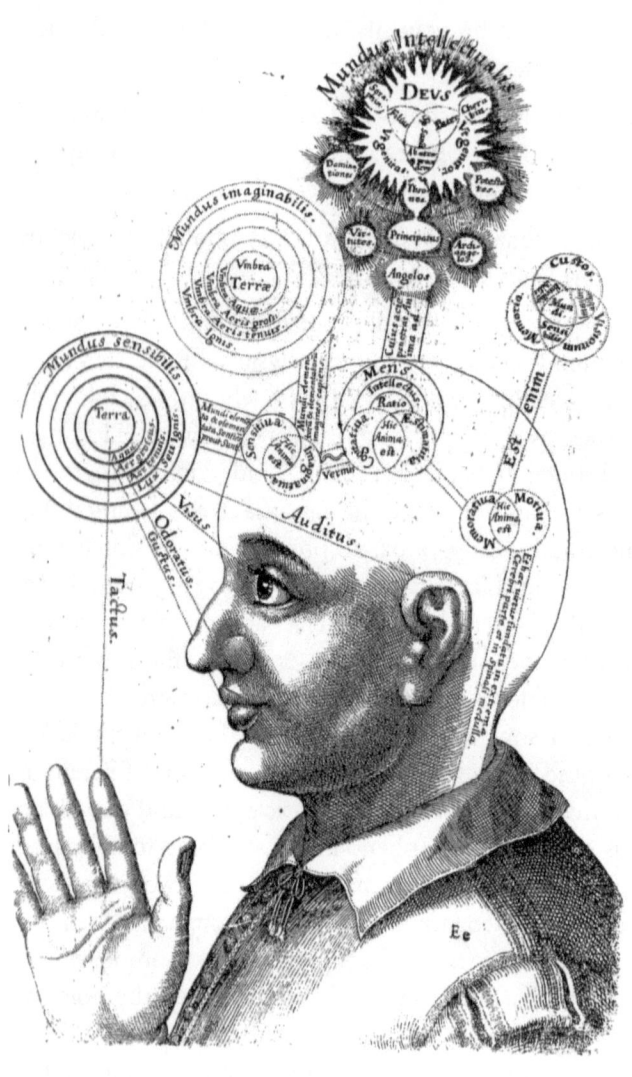

Figure 17: How human consciousness was attempted to be explained in the 17th century

"Please.

"I noticed your last phrase. They said that cognitive consciousness in its primary form consists of a 'representation of the environment'.

"Yeah?"

"In our previous topic we found that it is a conscious process that assigns a representation of reality to the pattern, and this representation is what we understand as meaning. If your statement is correct, my graph (Figure 16) confirms this and 'meaning' is part of cognitive awareness. - What do you think?"

Elisabeth Delacroix trims. She thinks for a moment before answering. "I think that the representation of reality is no different from a primary form of cognitive consciousness. On the other hand, cognitive consciousness also has more complex levels, such as reflexive consciousness. By this we mean the ability to follow one's own stream of thoughts. This is what I possess just by thinking about cognitive awareness. Cognitive consciousness is thus more comprehensive than a form of representation of reality. My remarks confirm your graph, which shows the meaning as part of a consciousness. Are you satisfied?"

"Thank you, I would be happy to continue with your original execution now."

Professor Delacroix continues: "The second differentiated type of consciousness is the phenomenal consciousness. It concerns the subjective aspect 'as it is' for example to feel pain or to see the colour blue. So it is about impressions and feelings of pleasant or unpleasant, which are difficult to describe. Therefore, as far as artificial systems or animals are concerned, we limit ourselves to the cognitive aspect of consciousness, which is accessible to the methods of neuroscience, experimental psychology and artificial intelligence".

The computer scientist Paul Aiken, who deals with questions of artificial intelligence, poses an interposed question: "I am particularly interested in which objective features make consciousness recognizable?

"Consciousness can be recognized by behavior that shows certain characteristics. Professor Allman, can you take down the following list right away?"

"Sure!" Professor Allman projects the list dictated by Professor Delacroix:

<u>Behavioural traits that indicate awareness</u>

1. the ability to discover changes in reality, to adapt to them and/or to pursue a goal, even if conditions have changed in a previously unknown, unexpected way

2. self-controlled behaviour that is not predictable with certainty and that uses information stored at some point in the past when required;

3. unpredictable, intentional behaviour which indicates the foreknowledge or anticipation of events and developments (anticipation)

4. the ability to critically evaluate one's own stream of thoughts.

If characteristics 1 to 2 are fulfilled simultaneously, this indicates primary consciousness. If in addition characteristic 3 or 4 is fulfilled, this indicates reflexive consciousness.

"What about examples?" Dr Maupertius the philosopher wants to know.

"We can examine two examples of the extent to which we detect consciousness there. The first example is the control system for human body temperature. For our purposes a highly simplified form of this control system is sufficient. In the second example we want to investigate whether chimpanzees or bonobos are conscious. - Professor Allman,

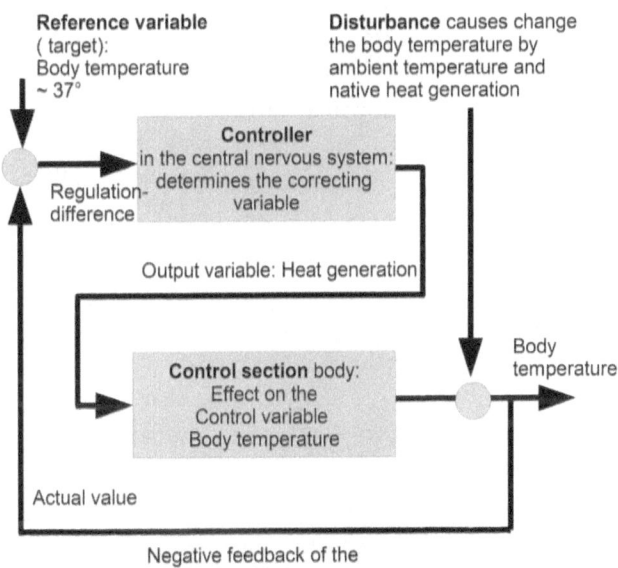

Figure 18: Body temperature control system

would you be so kind as to download and project the Body Temperature Control System diagram from the Internet?

Professor Allman is doing what he was asked to do.

"Let me briefly explain the diagram before we examine whether the regulatory system shows consciousness. Characteristic of a control system is the closed loop with negative feedback. The diagram shows the standard form of a control system as it is frequently found in biology, zoology and technology. Here it regulates the body temperature. It consists of the body as the control system, the central nervous system as the regulator and a negative feedback of the body temperature to the regulator. The body temperature is called the controlled variable (also: actual value). The system deviation is calculated from the difference between the setpoint and the controlled variable. The heat generated by the central nervous system (manipulated variable) affects the body (controlled system) and thus in turn the body temperature (controlled variable). As a disturbance variable, the ambient temperature causes a change in the body temperature (controlled variable), which is not desired and therefore has to be compensated via negative feedback. Nature has established a rather sophisticated system here, in which the central nervous system is involved. What do you think, dear colleagues: is the central nervous system the seat of primary consciousness, which manifests itself in this regulatory system for body temperature?

The participants puzzle for a while before Paul Aiken dares to formulate the answer. "The control system already possesses the ability, as required in characteristic 1, to adapt to certain changes in reality. Normal temperature fluctuations are among such changes. It is also clear that the control system pursues one goal, namely to keep the body temperature as constant as possible at 37° Celsius, which it normally succeeds in doing. However, if the environmental conditions change in a previously unknown, unexpected way, I have my doubts whether the control system can react appropriately. I try to imagine what the control system does when the body is in a 100° Celsius hot Finnish sauna. I don't think the control system can cope with the overheating for a long time. A real consciousness must intervene and instruct the body to leave the sauna as soon as possible. Furthermore, the control system will not store information about how hot a Finnish sauna can get and will use this knowledge for its future control behaviour. Neither characteristic 1 nor characteristic 2 is fulfilled. Primary consciousness is therefore not recognizable."

"Thank you, Mr. Aiken. That was exhaustively answered. Emotionally, it was clear that the system of regulating body temperature could not show primary consciousness, but without objective features the reasoning would have been vague. Therefore it is a beautiful example of the application of the characteristics of consciousness to real systems. Now the second example. Professor Allman, would you please download my film about the experiments with the chimpanzees from the Internet and show it to us?"

"Nothing would please me more!" A few minutes later the film is shown.

The movie title pops up: "Spot Test After Gordon Gallup". The following scenes show chimpanzees in a species-appropriate attitude. Then an animal psychologist is introduced and shown how he prepares a banana with a sleeping pill. Afterwards he approaches a female monkey, whom he addresses in a friendly way with Lucy. He hands her the banana I prepared. Lucy eats these and falls asleep a short time later. The psychologist takes Lucy in, carries her to a

Figure 19: Do common chimpanzees (Pan troglodytes) possess consciousness?

special dwelling and places her in front of a large mirror. Before he leaves, he paints a red spot in the middle of her forehead with a body paint.

When Lucy wakes up, she watches her counterpart in the mirror attentively. She needs a few seconds before she realizes that something is not normal. Then she grabs her own forehead with her right hand and tries to remove the red spot.

Again and again she scratches the stain with a finger and observes the emerging success in the mirror.

The movie ends. The seminar room is filled with thoughtful silence.

"Is Lucy conscious or not and why?" asks Professor Delacroix with a smile.

Dr. Dessoir is trying his luck. "The first thing I notice is that Lucy was not examining her reflection, but herself. Do all monkeys act like that?"

"No, not everyone behaves like that, but chimpanzees know that the mirror reflects their own image."

"You obviously also know that The Blur is not normal."

"That's right!"

"Based on the characteristics of consciousness set forth earlier, I would say Lucy has the ability to detect and adapt to changes in reality. The red spot is an unexpected, probably previously unknown way of change. I would therefore like to consider behavioural criterion one to be fulfilled."

"That's right too!" comments Professor Delacroix.

"Regarding behavioral trait 2, I can say that Lucy was probably not trained to remove red blotches from her forehead. This means that it shows unpredictable, self-directed behaviour. To remove the stain, she uses her life experience, which she has stored sometime before. Therefore, for me, behavioural criterion 2 is also fulfilled. That means Lucy at least shows primary consciousness."

"You have correctly assessed Lucy's behavior, Dr. Dessoir. She's conscious!"

"Thank you! I'm just puzzled that in my opinion Lucy shows more than primary consciousness when she doesn't treat her reflection as another individual, but recognizes herself in it. On the other hand, I have difficulties in applying behavioural traits 3 or 4. What do you think, Professor Delacroix?"

"The stain test is considered proof of self-awareness. I-consciousness belongs to the reflexive consciousness. The behavioural characteristics 3 and 4 are really difficult to assign in the example. The stain test should actually be included in the list of behavioural characteristics as a further behavioural characteristic, namely number 5. I only left it out because I can use it as an example of primary consciousness. - But now I would like to turn the floor back over to Professor Allman."

"Thank you very much, Madam Colleague! - They have taken us so far that we can now apply the behavioral traits that indicate consciousness to experiments with quanta."

"Not enough for animals to have consciousness. You don't want to prove consciousness in quanta," theologian Johanna Balthasar says.

"Why not, if it were possible?" wonders Professor Allman.

"Because it is sacrilege!"

Professor Allman explains his attitude in a calm tone. "I am aware that until a few centuries ago people were burned at the stake when they questioned the prevailing doctrine of the time and claimed that the earth was neither a disc nor that the sun orbited around the earth. In the meantime, there is no longer a monopoly on answering naturalscientific questions. Free, critical thinking is no longer threatened with death, at least in our country. The majority of the colleagues present here will agree with me and are certainly eager to see what the application of the behavioural traits to quantum experiments will show."

"Let me try to apply the behavioral traits!" Professor Geiger intervenes. "The first question is: Can a quantum - within the limits of its possibilities, of course - detect changes in reality and adapt to them? In the double-slit experiment it immediately recognizes when a gap is covered. Then the quanta know that they can look again for the dark stripes they left out during the interference. Even if both columns are open and they notice the intention of the experimenter who wants to find out which way they are going, they immediately change their behaviour and do not interfere anymore. They apparently do not want to disclose the routeinformation under any circumstances. - In my opinion, quanta are not about pursuing a goal. Nevertheless, as a physicist one has the feeling that their goal is to withhold certain information from us. No matter how we change the experimental conditions in order to get at them, they adapt to the changed conditions without exception and refuse to inform us how an individual one of them behaves, because after each measurement it independently selects a real state that is not predictable. I therefore consider criterion one to be met."

"And what about the predictability of behaviour and self-control and the use of previously stored information, i.e. what is important for behavioural criterion 2?

"Let me first answer the question about the use of previously stored information. The cosmic variant of the double-slit experiment shows that even after billions of light years or after detours of 50,000 light years, quanta still know that they belong together and come from the same light source. They can cope with changes of the direct path - this again confirms criterion 1 - on the other hand it shows: They have stored the information about their origin, no matter what adverse circumstances their path may have, they interfere! - When it comes to unpredictable behavior and self-control, the twophoton experiment is the proof. It is absolutely unpredictable which path a single quantum will take at the polarizing beam splitter PBS, no matter what experiments we physicists do to get to know the quantum's decision beforehand. It unpredictably selects either the horizontal or the vertical polarization and then changes to a real state. There are many other experiments with quanta, which cannot all be listed in this course. The experiments have one thing in common. The quantum always seems to choose its real state itself. The choice can never be foreseen. I therefore consider behaviour criterion 2 to be fulfilled. This surprises me all the more because before this seminar I would never have thought that quanta could possess primary consciousness!

Professor Allman shakes his head: "Don't be angry with me, but I think that quanta do not possess primary consciousness, even if they have argued well, Professor Geiger!

"I must disagree with you," Elisabeth Delacroix enters the discussion. "If I take my own list of objective behavioral traits for consciousness seriously, then I must agree with Professor Geiger!"

Things are getting restless in the hall.

Paul Aiken expresses his opinion with great commitment: "I share the same opinion as professors Geiger and Delacroix!

"Thank you, thank you, dear colleagues, for your committed opinions! - You're about to see that I'm still right. I didn't say that quanta would not show primary consciousness through their behavior, I just said that they would not possess one. There is a difference!"

Paul Aiken calms down. "You'll have to explain that, Professor Allman."

"There are two arguments. The first is Plato's Parable of the Cave. The shadow that shows consciousness is not reality. We must first work for reality. And the second argument is one from information theory. I'm asking you for a professional valuation, Mr. Aiken. Suppose you could develop a software program that shows conscious behavior, how many bits would such a program contain?"

"Oh dear, that's very hard to estimate! - I think, in order to realize the simplest approaches of primary consciousness programmatically, one would need many megabits of program code".

Professor Allman turns to Professor Geiger. "Please tell us how many quantum bits a single quantum can store?"

"It varies widely, from less than eight to several hundred!"

"This answers whether a single quantum can have consciousness. Even primary consciousness requires many more bits of information than a quantum can have. Therefore a quantum can show consciousness but not have it."

"Amazing!" Elisabeth Delacroix brushes her curly hair strands from her face. "You must explain to us where the consciousness that the quanta show is located, Professor Allman!"

"The only possibility outside our space-time universe is vacuum. This complements the properties of the vacuum". Professor Allman completes the corresponding slide.

Properties of the vacuum
- topological **space without metrics** (distances do not matter)
- the points in space are **information memories**...
- ... contain the **vacuum energy**
- ... and possess primary **consciousness**

The philosopher Dr. Maupertius has listened very carefully, but he is not yet completely convinced. "I see: The vacuum is the carrier of the primary consciousness. But I don't understand how quanta manage to show primary consciousness that they themselves do not possess.

"In response, I would like to show you a small physical experiment that can serve as an example of the interaction between energetic objects." Professor Allman starts the short film, which shows a shot-put pendulum in action.

Five balls are suspended from threads and these are fixed in a holder. The ball, which is located far left outside, swings to the left, stops at the highest point and swings back. It collides with a row of four stationary balls, which are also suspended from pendulum threads. At once the left ball stops and remains in peace. Ball outside right side swings out instead. When it swings back and hits the resting balls, the leftmost ball swings out again. The game begins again (Figure 20).

Professor Allman begins to explain. "What interests us is the moment of interaction between the spheres. The interaction is only very short and takes place at the moment

Figure 20: The shot-put pendulum illustrates a very short interaction with exchange of energy and information, Photo: Dominique Toussaint, GNU license see appendix

when a ball swings back and hits the resting balls. At this moment it transfers energy to the resting balls. We now know that a pattern can be stored in energy, which belongs to a certain information. Here it is the information about the speed and the direction of oscillation. The information is passed on together with the energy to the next sphere. In a further interaction, the latter passes both on again. The last ball on the right has no partner to pass on to. Therefore, it carries out what energy and the associated information is applied to it, namely to oscillate at a certain speed in a given direction".

Dr Maupertius still sees no connection with his question. "That's all well and good, Professor Allman, but how is the primary consciousness of vacuum transferred to quanta?"

"Consciousness is not transferred at all, Dr Maupertius. It is information that is transmitted, namely the information that the quanta need to show their specific behaviour as manifested in the experiments. If you look again at the picture that shows information as part of consciousness, you should be able to see the connection. The consciousness remains in a vacuum. The interaction with the quanta occurs like any known interaction by passing on information. In this case it is information that indicates consciousness. And that interactions between the quantum and the vacuum must actually take place, I have proved to you by the otherwise inexplicable phenomena."

"Thank you, Professor Allman, it's just too much new to have everything in one's head right now. I will go over my notes again tonight to recap what I have already said."

The parapsychologist Dr. Dessoir makes a sour face: "Have you forgotten my request, Professor Allman? - In connection with your vacuumtheory you had promised that you would prove that there is *a transpersonal consciousness whose origin lies in the vacuum.*"

"Wait a minute, Dr. Dessoir, actually, I thought But I'd like to take care of it. Before doing so, I must obtain a professional information from my esteemed colleague Professor Delacroix."

"Yes, please, what is it, Professor Allman?" Elisabeth Delacroix smiles.

"Just one question: As a consciousness researcher you will certainly know where human consciousness is generated. Can you tell me?"

"Until recently I would have replied: in the human brain! - This course, Professor Allman, has made me unsure. Although brain function correlates with consciousness, nothing proves that it actually generates consciousness. This is like the quanta showing consciousness, although it is only an interaction with the primary consciousness of the vacuum. I must therefore say that where human consciousness is generated has become an inexplicable phenomenon for me."

"Thank you, Professor Delacroix! - Now for my argument. To explain the phenomenon of transpersonal consciousness, we need the continuum model of Dr. Anaximenes. Since there is no local connection between the consciousness of the sender and the receiver, there must be a non-local, metric-free connection via the vacuum. However, this phenomenon cannot be compared with quantumphenomena, as quantumphenomena at the RZU always have a local connection between the transmitter (light source) and the receiver (measuring device). On the other hand, transpersonal consciousness can only be explained if a similar explanation is assumed, as is the case with quanta that show primary consciousness. This means that the brain functions of the transmitter and receiver show consciousness, but **in reality at least parts of the human consciousness belong to the vacuum**".

Professor Allman pauses at this point to let his words sink in.

Some participants are breathing heavily. Someone in the back row quietly says, "This is getting great!"

Professor Allman hears it and gives an answer: "Yes, that's how it is! Once we discuss the true face of reality, I am afraid that your previous world view will be in ruins and replaced by a new one. But let me first finish the reflections on human consciousness before we start a new topic!"

Elisabeth Delacroix takes the opportunity to contribute a thought of her own. "It makes sense to me! Why should the space points of the vacuum be limited to primary consciousness when they have been proven to already possess consciousness? - Since, in contrast to photon experiments, there is no local connection, the connection between the consciousness of the human transmitter and that of the receiver must be made exclusively via the vacuum. From this follows an interaction between human consciousness and the space points in the vacuum, which have at least primary consciousness. Nothing proves that brain function actually generates consciousness. This can be explained by assuming that consciousness is created in a vacuum. I think **the brain function only shows consciousness, which is actually a function of the vacuum.**"

The parapsychologist Dr. Dessoir becomes increasingly agitated while Elisabeth Delacroix presents her line of argument. His right hand brushes across his hair and forehead. Finally, he blurts out. "There are many reports of near-death experiences with unexplained phenomena. If one assumes that consciousness is not a brain function but a function of the vacuum, the phenomena could be explained."

"Is that the kind of case you want to present to us, Dr. Dessoir?" asks Professor Allman.

The excitement leaves Dr. Dessoir as he begins to tell: "There is even a BBC documentary film called 'Encounter with Death' from 2003, in which a well-known American songwriter by the name of Pam Reynolds describes her experiences during her death-like condition during surgery for an artery dilation in her brain. She got severe dizziness and could no longer speak. The artery dilation was at the base of the brain, so that regular surgical methods could not reach it. The neurologist examining her did not give her any hope of a cure: the dilatation of the artery was also a time bomb that would certainly burst soon. She'd have to die anyway. As a last resort, Pam Reynolds decided to confide in the neurosurgeon Dr. Robert Spetzler. Spetzler has experience with the surgical procedure 'hypothermic cardiac arrest'. The patient's body is artificially hypothermic, causing cardiac arrest and thus a death-like state. This makes it possible to operate on parts of the brain that are difficult to access. - Do you want to download the film documentation from the Internet and show it here, Professor Allman?"

"Gladly, but it'll take a few minutes. In the meantime, you could elaborate on the surgical procedure."

"The patient is anesthetized and gets his eyes taped. Special earphones are inserted into his ears for click tests to measure brain waves. The earphones prevent the patient from hearing anything other than clicks. For this reason alone he can neither see nor hear anything that is happening around him. - Normally the human brain tolerates the interruption of the oxygen supply for a maximum of three minutes. Afterwards it dies off. However, due to severe hypothermia of the body it tolerates the interruption for more than an hour. A modified heart-lung machine is used to achieve the hypothermia necessary for the operation at the base of the brain. The blood is cooled in a heat exchanger outside the body, in Pam Reynolds' case to 15.5 degrees Celsius. Once the desired temperature is reached, the surgeon interrupts the heartbeat with an ice-cold potassium chloride solution. Respiration and brain waves stop, the measuring instruments show a zero line EEG. The metabolic activity of the brain is stopped. Every registerable manifestation of life of the body is stopped. This one is in a deathlike state."

Professor Allman has loaded the film documentation and fast-forwards until the scenes with Pam Reynolds appear.

> *Pam Reynolds says, "I don't remember an operating room.*

Dr Dessoir comments: "So she was no longer conscious when she was brought into the operating room."

> *Pam Reynolds: [...] I don't remember anything until this sound. It reminded me of a dental practice. [...] I*

remember the top of my head kind of popped out. And then I saw a body. It was my body. [...] My vantage point was somehow on the doctor's shoulder. I remember the instrument in his hand. It looked like the handle of my electric toothbrush. I had assumed that they would open the skull with a saw [...] but what I saw here reminded me more of a drill. There were even several small drills in a box.

Dr. Dessoir: "Later in the documentation, Dr. Spetzler will testify that Pam Reynolds could not see the equipment used when she was brought into the operating room, even though she may not have been fully asleep. The devices always remain covered or packed until they are used. This is necessary to ensure sterility. When they are used, the patient has long since fallen asleep completely. - I myself think her eyes were glued shut, how could she have seen anything? - Even in her imagination, she could hardly imagine the appearance of the saw resembling a toothbrush, since even commenting experts, such as the following Dr. Michael Sabom, had no idea what a bone saw for opening the skull cap would look like. Dr. Sabom had a photo sent to him by the manufacturer of the device, and only then did he realize that the saw actually resembled an electric toothbrush.

Pam Reynolds continues: "And I clearly remember hearing a woman's voice say, 'we have a problem, her arteries are too narrow'. It seemed to come from further down the table. I wondered about that because it was brain surgery after all. But they had placed an access to the femoral arteries to drain the blood."

Dr. Dessoir: "There was no way to hear conversations about her hearing because her ears were plugged! - Dr. Sabom analyzed the surgical protocols and spoke with Dr. Spetzler, the chief surgeon. He says that Reynolds' testimony was very close to what actually happened and that she could precisely recall a conversation between Dr. Spetzler and the cardiovascular surgeon who had to open an artery to connect it to the heart-lung machine. - I repeat, at this stage of the operation no patient can see or hear anything!"

> *Pam Reynolds: "[...] I saw the pinhead-sized light. The light began to attract me. I walked towards the light. The closer I came to the light, the more I recognized different people and I clearly heard my grandmother calling for me. I went to see her immediately. It was a great feeling. [...]"*

"Less interesting for our naturescience course," murmurs Dr. Dessoir, and continues louder: "Would you fast-forward the film a little, Professor Allman?"

Professor Allman is doing what he was asked to do.

> *Pam Reynolds: "At some point I was reminded that it was time to go back. "I saw the jump into the body...*

Professor Allman stops the film documentary. "I think we've seen enough to discuss." His gaze looks for hand signals from participants who want to say something. Finally, his gaze falls on Dr. Alfred Schleich, an anesthesiologist of the skeptic movement. He's the only one calling in at the moment.

Figure 21: Hyronimus Bosch; The Flight to Heaven; around 1500; Location: Doge's Palace in Venice

Reluctantly, Professor Allman asks, "Yes, Dr. Schleich, what is it you want to say?"

"I belong to the Association of Skeptics. We have made it our mission to expose all pseudo and parascientific theories in a completely unbiased manner. Based on my 20 years of professional experience as an anaesthetist, I believe that the patient was conscious several times during the operation. Their impressions came about by mixing actual perceptions with the effects of the strong medication. It is therefore a phenomenon that can be explained quite normally, but which cannot contribute anything to a vacuum theory here in this course. We should therefore move on to the next topic."

Professor Allman ignores the professional sceptic's request to change the subject and hooks in elsewhere. "I have heard from your words that you do not doubt the documented facts. You would like to see the facts explained only as a normal process. Is that correct, Dr. Schleich?"

"That's right, Professor Allman!"

"In her opinion, the patient had at least partially conscious perceptions of her surroundings. Other impressions were caused by the strong medication?"

"Yeah, sure. - On superficial observation, the story seems like proof of life after death. But it proves nothing!"

"Thank you for your contribution, Dr. Schleich. I am also of the opinion that the facts of the documentation are not sufficient to prove directly and without further arguments, a life after death. What makes the case interesting, however, is the fact that the story happened under medical supervision during a fully logged operation. Therefore certain facts cannot simply be discussed away. What are the facts?"

Professor Allman received requests to speak. From these he forms two statements, which he projects into space:

- Conscious, optical perception of the surgical instruments despite closed eyes
- Conscious, acoustic perception of the conversations of the surgical team despite soundproof plugged ears

"So there was an interaction between what happened at the RZU, i.e. the operating theatre, and the patient's consciousness," summarises Professor Allman and then asks the crucial question: "Is it a local interaction like watching a television film from an armchair?

"No, it certainly wasn't a local interaction!" Dr. Dessoir replies and shakes his head.

"And why not?" is what Professor Allman wants to know.

"This is clear: only the sensory organs could have mediated a local interaction between the external events and the patient's consciousness. But the patient could neither see nor hear anything with her sensory organs."

Professor Allman is satisfied: "All right, if it was not a local interaction, then the only alternative is a non-local interaction. - By the way, do you remember where we had already discussed a non-local interaction during the course?"

"Thespooky long-range effect of the two-photon experiment," shouts someone from the back rows.

"Right!" says Professor Allman. "We have also set up a mathematical model to illustrate the relationship between local and non-local interactions. Who remembers what model it was?"

"The continuum model Figure 13!", echoes back from the participants.

"Good! then take another close look at this continuum model and then answer the following question: Where must the origin of the patient's consciousness lie so that it can enter into a non-local interaction with the RZU, i.e. the operating theatre?

The room is rustling. The participants leaf through their documents. A short time later they discuss quietly with each other. Little by little, an opinion prevails. Dr Dessoir reporting. At Professor Allman's request, he reports.

"We believe that the origin of the patient's consciousness is the vacuum!"

"Can you justify this?"

"Yes! - There are two theories. According to the first theory, the brain of the patient is the origin of consciousness. For an interaction between the two locations A and B in the RZU, two non-local connections, i.e. interactions with the vacuum, are required. - According to the second theory, the vacuum is the origin of consciousness. In this case only one interaction is required during the operation, namely between the RZU and the vacuum. Because according to the scientific principle of 'Ockham's razor' with two theories that explain the same facts, the simpler one is preferable, we have decided to go for the second theory. **The origin of consciousness is the vacuum.** This, by the way, fits in with the knowledge we had already gained in the course before and once again confirms its correctness.

"Wonderful!" says a delighted Professor Allman. "So during an operation, the origin of consciousness is the vacuum. After surgery, when the patient wakes up. ...does the consciousness move from its point of origin into the patient's brain?"

Dr. Dessoir is baffled. Then he must laugh. "No, certainly not! Ockham's razor can also be used here. The simplest explanation is the right one! Consciousness remains where it is, namely in a vacuum. The brain practically serves as a receiver. It shows consciousness but is not its origin."

"There is not much I can add to that," confirms Professor Allman. "However, before we move on to the next topic, I would like to use what we have learned about the vacuum and consciousness to reformulate the properties. While typing on the computer keyboard and speaking to it, the beamer throws the text into the room:

<u>insights into consciousness</u>
- We know that the space points of the vacuum are information stores. The information storage is not realized by matter, but by a type of energy. This type of energy is the vacuum energy. The information of the memory is part of the consciousness. Compare Figure 16: Consciousness is therefore the comprehensive, the primary. It contains energy and information. As it does not contain anything else according to our knowledge, it can be equated with energy and information. It follows: **Consciousness is a type of energy**.
- The vacuum is a topological space without metrics (distances do not matter). The space points of the vacuum contain consciousness units of different sizes. There are small units of primary consciousness related to quantum effects or the certainly larger units that can be attributed to human consciousness.

- The units of consciousness are connected to each other and to the events in the space-time universe by a metric-free, i.e. non-local interaction.

"In summary, information may be the fundamental building block of the space-time universe, but I would like to share another important insight with you at the end of the day.

Professor Allman handwrites the following sentence on a slide and projects it:

> **CONSCIOUSNESS IS THE FUNDAMENTAL BUILDING BLOCK OF EVERYTHING THAT EXISTS.**

"The realization follows firstly from the fact that consciousness contains the information from which the space-time universe is built. Secondly, it follows from the fact that the energy of the vacuum is responsible for the creation of matter. Just think of Einstein's famous formula about the equivalence of energy and mass! This energy must be equated with consciousness, since consciousness is the energy type of the vacuum. With the concluding words: "Get a good rest so that you can participate tomorrow in the old freshness again", he says goodbye.

The true face of reality

The secret is that in every moment
I am more and more in tune with each other and yet
always the same.
That I am always the same,
my consciousness causes; that I
am different in every moment,
causes space and time.

Leo N. Tolstoy, Diaries (1906)

Thursday, June 5th

The philosopher Dr. Maupertius supports his head with his hand as if he were still tired from last night. "What I would like to know, Professor Allman, is whether yesterday's new knowledge changes our conception of space and time.

"Yes, quite enormous, I can only say, Dr Maupertius! - You will see that time is only an illusion and the same applies to the viewing room as we experience it every day. But let's take it slow. For the non-physicists among those present, I would like to address some essential points of our current conception of space and time before I show you what the true face of reality looks like. - Professor Geiger, can you tell us something about the Michelson-Morley experiment, whose unexpected outcome before 1905 was considered an inexplicable phenomenon?"

"Gladly! - The Michelson-Morley experiment is one of the most important experiments in the history of physics," answers Professor Geiger not without pride. "Michelson first performed it in 1881, Morley then in a more refined form around 1887. At that time it was assumed that space was filled with ether. The experiment should measure the relative speed with which the earth moves through the ether. Michelson and Morley believed that the Earth would behave in a similar way in the ether, like a football in the air, and produce a detectable 'etheric wind'. - Imagine a swimmer in a river! - Effect of ether wind onto light waves should be just like effect of flux onto this swimmer, who once swims upstream and downstream. Upstream the swimmer is slower than downstream. With the help of an ingenious experimental setup, the Michelson interferometer, it was therefore only necessary to measure the speed of light both in the direction of the ether wind and perpendicular to it. By difference of both speeds one wanted to calculate ether wind.

"And what was the result, Professor Geiger?" asks Professor Allman with an expectant smile.

"The result of the measurement was a big surprise. No matter what you did, no matter in which direction you measured the speed of light, there was no difference. The speed of light was the same in every direction. Although the earth moves around the sun at about 30 km/second and therefore the speed of light in the direction of the orbit should have been different from the speed of light in the opposite direction, no difference could be detected. One believed in measurement errors and repeated the measurements again and again. Always the same result: **The speed of light is constant!** "

"It was an inexplicable phenomenon at the time. Who and when did it lead to an explanation?"

"Albert Einstein published his special theory of relativity in 1905. This could explain the result. The old ether hypothesis became obsolete."

"What is the significance of relativity?"

"The theory of relativity has revolutionized the understanding of the world. The special theory of relativity describes how space and time behave from the point of view of observers moving relative to each other. - Relativity theory and quantum theory are today the two pillars of the theory building of physics. Unfortunately, it has not yet been possible to unite the two theories into one."

"There's a good chance that my theories on vacuum and consciousness will lead to this union," smiles Professor Allman mischievously. "What are the statements of special relativity?"

"We must abandon intuitive notions of absolute space and time. Space and time are not universally valid in the theory of relativity. Spatial and temporal distances of events and thus also their simultaneity are judged differently by observers with different states of motion, without it being possible to say that one of these observers is right. Moving objects appear to the resting observer to be shortened in the direction of movement, and time seems to pass more slowly. The first phenomenon is known as length contraction, the second as time dilation."

"This is all just grey theory or science fiction ...", Johanna Balthasar interrupts annoyingly, "and has no meaning whatsoever in everyday life! The speeders on the freeway don't seem shortened to me at all, just their minds."

Some participants feel exhilarated and laugh softly.

"Make no mistake," Professor Geiger replies calmly. "I'm sure Professor Allman will soon bring proof where relativity affects our everyday lives!"

"Yes, right away!" replies Professor Allman. "But first I would like to show by means of diagrams why lengths are shortened in moving objects and why time passes more slowly." He projected a Minkowskid diagram in the manner developed by Hermann Minkowski in 1908 to illustrate the properties of space and time in special relativity.

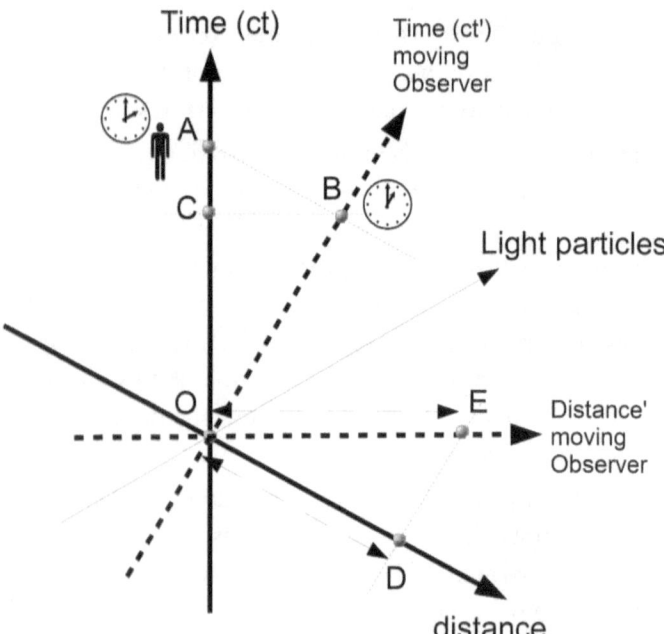

Figure 22: Minkowski diagram illustrating time dilation and length contraction. For the observer in A the moving clock in B seems to run slower. The length OD determined by him for the moving scale OE is shorter.

Professor Allman explains: "The basic diagram is a simple path-time diagram with the two axes distance and time. In order to be better suited to represent the conditions in general relativity, the time ct multiplied by the speed of light c is used as the time axis. A light particle, which always has the constant speed of light c, moves on the bisector between the two axes. An observer A, at rest with respect to this coordinate system, does not move away from the zero point in the direction of the distance axis, but in the direction of the time axis as time progresses. The line of a moving observer B is the dashed axis lying between the axis of the light particle and the time axis ct".

"Why did you draw another dotted line of distance, Professor Allman? In a normal path-time diagram there is only one single distance axis," consciousness researcher Professor Delacroix objects.

"As has already been stated, the speed of light is always constant, even for the moving observer. If I draw a light particle on the bisector between the two axes for the observer at rest, then the same light particle must also lie on the bisector for the moving observer. This is, as can be easily seen, only possible if the moving observer has his own distance axis, namely the dotted one. - A question of understanding to you, ladies and gentlemen: On which straight line must all events lie that the observer in A interprets as simultaneous?

Dr. Krates is faster than the other listeners: "On a parallel to the distance axis through point A! - But why did you also draw the moving observer on these parallels? "

"Because, from A's point of view, this illustrates the slower-running clock in the moving system! - Compare the time period OA with OB! Which is smaller?"

Dr. Krates need not think twice: "The time distance OB is shorter!"

"There you have it! - The clock of the moving observer seems to run slower for A! - This is the time dilation! - Conversely, B believes that the clock of the allegedly resting observer has only arrived at C, i.e. it is the clock that is slowing down. Everyone thinks the clocks of the others went slower. Which of the two is right, one cannot decide. There's no point in asking."

"Are points D and E also connected with such a paradoxical situation?" Elisabeth Delacroix would like to know.

"That's right, it's all about the contraction in length. But here I have only entered a contraction for reasons of simplification. A moving observer B carries a scale of length OE with him. For the allegedly resting observer A has the smaller length OD."

Johanna Balthasar argues: "I can only repeat: All grey theory that has nothing to do with reality!"

"No way! - The following picture proves the importance of the theory of relativity for our everyday life," answers Professor Allman calmly and thoughtfully. It projects the image of a GPS satellite needed for positioning (Figure 23). "Who can tell me why the theory of relativity is being touched on here?"

The computer scientist Paul Aiken knows the answer: "GPS satellites constantly send their changing position and the time of their extremely accurate atomic clocks. GPS receivers on earth can use this information to calculate their own position. Now satellites are moving objects that orbit the earth at high speed. Although the atomic clocks in the satellite are

Figure 23: Second generation NAVSTAR satellite for positioning

extremely accurate, they run slower than the same clocks on earth due to time dilation. The transmitted GPS data would become useless in the shortest possible time if they were not corrected using the formulas of relativity".

"This is empirical proof of time dilation," says Professor Allman. "There is also empirical evidence for the length contraction. They found muons at sea level. The very short-lived and almost light-fast muons, which are generated in the upper layers of the atmosphere by the impact of cosmic rays, should have decayed long before they reach the sea surface. But because of the considerable contraction in length close to the speed of light, the distance to the earth's surface seems to be much shorter. That's why these muons are actually still found at sea level."

"You should mention the concept of space-time at this point, colleague," says Professor Geiger in a good-natured manner.

"Good! - In a three-dimensional world, time dilation and length contraction would be impossible. For this, the world must be four-dimensional and must have a time dimension in addition to the three spatial dimensions. Now, however, it is the case that space and time appear to be of equal value side by side in the basic equations of relativity. There is no essential difference between space and time. This is why physicists have coined the term **spacetime** for our four-dimensional world. We do not want to say space-time-universe or RZU in the future, but space-time or space-time-universe. I mean, space-time forces us to have a completely new idea of how objects look in four-dimensional space. I now deviate from the currently still prevailing view and would like to illustrate my view of things with two diagrams".

"The left diagram (Figure 24) shows the conditions in three-dimensional space. It is a normal path-time diagram for a cube-shaped object. Time does not belong to space, therefore the diagram represents the **temporal development**. - The

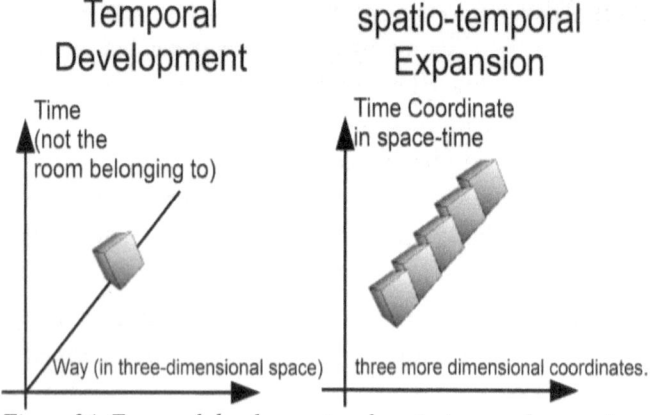

Figure 24: Temporal development and spatio-temporal expansion

diagram on the right is intended to illustrate the conditions in four-dimensional space. The time coordinate here belongs to space. There is therefore no temporal development, but rather a **spatio-temporal expansion**. I want you to understand the difference. - In the three-dimensional viewing room there is also no development of two-dimensional cube surfaces. The cube is three-dimensionally complete in its expansion. Just as complete in its four-dimensional extension must be the cube and of course every object in the four-dimensional space-time.

Dr. Krates swings his head back and forth: "How is it then that we are not aware of the spatio-temporal extension of such a cube? - We see the cube rather in its temporal development."

"Have a little patience, Dr. Krates! I would like to finish the previous one first. - **Spatio-temporal expansion means that an object contains all its eigenstates."**

The physicist Dr. Helmholtz is startled as if electrified: "That's true of quanta! These contain all their eigenstates before they are measured, decide on one and then become real. I'm thinking of the vertically and horizontally polarized photons, which contain both states before they have to choose one in the PBS beam splitter".

"With quanta, the spatio-temporal expansion before decoherence is quite obvious. But even after decoherence they still have spatio-temporal extension, even if this is no longer so obvious. And, of course, all objects are spatio-temporally extended."

"Are there other insightful examples by which the spatio-temporal expansion can be easily recognized," asks Edward Michelson.

Professor Allman ponders briefly before answering: "A film strip is in a sense a spatiotemporal extended image, because the film strip can be interpreted as an object that contains all the intrinsic states of an image.

"What are the consequences if all objects are spatiotemporally extended?" Michelson continues.

"Seen as a whole, the objects are static in their spatiotemporal extension. There is no development over time. Time as a flowing something does not exist. **Time is an illusion**. For example, a film strip as a whole is static. There's nothing moving!"

Dr. Krates stubbornly returns to his previous question. However, he reformulates. "Then I'd like to know why you move in front of us, Professor Allman?"

"Because consciousness comes into play, Dr. Krates, namely consciousness with a three-dimensional focus on the space-time universe. - The special theory of relativity presupposes consciousness in the form of the conscious observer moving at a constant speed. This ominous observer has influenced the formulas. Without the observer, the theory of relativity would be meaningless and physics would lose one of its two pillars. She might stagger and fall on her face."

"Heaven forbid that should happen," mumbles Professor Geiger in horror.

Professor Allman has heard it: "Don't worry, Professor Geiger, the theory of relativity will remain with us. We have seen before that consciousness is not a function of our brain. Thus it is also not statically anchored in four-dimensional space-time. As a function of the vacuum, it lies outside. Via a receiver within space-time, namely the brain, it interacts with the objects and has its three-dimensional focus on them. The focus changes in a specific sequence. The consciousness interprets the sequence as flowing time. This can be imagined similar to a film strip that is shown. In one moment you see only one picture, followed by the next and so on. Thus life comes into an otherwise static world!"

Dr. Krates remains skeptical: "Doesn't a static world mean that past, present and future are fixed? - Consciousness would then only be the observer of an unchanging film."

Professor Allman smiles: "To stick with the parable of film: If the consciousness had no free choice and could not choose the film, then it would indeed be the case that the present and the future are fixed in addition to the past. There would only be one big movie."

"Isn't it?"

"Have you ever heard of Many-worlds Theory, Dr. Krates?"

"No, that's not part of my philosophy department!"

"Well, let a physicist explain it to you!"

"I'll be glad to do that," says Dr Helmholtz. "Roughly speaking, according to the Many-worlds Theory, the space-time universe splits into two when a measurement is made or an observation takes place. Remember the PBS beam splitter in the two-photon experiment? - The photon must decide at the beam splitter whether it wants to come out vertically or horizontally polarized. The result cannot be predicted. It's pure coincidence. According to the many-worlds theory, the fission creates a universe in which the photon emerges vertically and a second one in which it emerges horizontally polarized. Both universes are in some ways parallel, coexisting realities. The observer is also split. Every observer sees only one reality, of course. Since many such processes take place continuously, in which a quantum must decide, there is an enormous amount of simultaneously existing universes.

"Who would have thought of such a thing?" wonders Dr. Krates.

"The theory is an interpretation of quantum mechanics and its experiments, such as the two-photon experiment. It goes back to Hugh Everetts. The expression 'many worlds' was coined by Bryce DeWitt, who covered the subject in more detail than Everett's. Mathematically and physically, the Many-worlds Theory is simpler than all other explanations of the quantum mechanical phenomenon of decoherence, i.e. the process by which quanta assume a real state. According to the principle of Occam's razor, the many-worlds theory is therefore preferable".

"Good, so there is not only one great film, but an infinite number!", Dr. Krates sums up. "But can the consciousness be free to choose a film and do it again and again?"

Professor Allman nods. "This will be our topic tomorrow! - There's enough stress on the head for today. Have a nice evening."

Answers to basic questions of our being

> *Being is completely first recognized in its scope and inner being as something that has become.*

Alexander von Humboldt, cosmos

Friday, June 6 - last day

Free will

The theologian Dr. Aniane leafs through his documents. Suddenly he turns pale: "I knew it, we have no free will!"

Professor Allman looks at him in amazement: "Even though the opposite is already apparent in quantum theory, I would be interested to know what makes you think so, Dr. Aniane?

Dr. Aniane hectically utters sentences: "You said yourself that the world is static. With spatio-temporal expansion nothing moves anymore. Time is an illusion, which consciousness makes us believe! Past, present and future are fixed in a space-time universe."

"Easy, easy, Dr. Aniane! - Calm down first and do not throw the child out with the bath water! - The past is past, that's right. A four-dimensional space-time is something static, that is also true! But every consciousness has a free will and determines its own future.

"I can't believe this! Not after all you've proven to us so far, Professor Allman."

"Well, let's start with the concept of free will. What does free will mean? I'd like to hear a physicist's answer to that question. Professor Allman looks around.

Enrico Fechner feels addressed. "If decisions cannot be predicted with certainty, they can be seen as the manifestation of free will. Everything that can be foreseen with certainty, on the other hand, is determined and cannot be the expression of free will. Wherever we physicists have a formula for reliable predictions, free will does not prevail!

"All right Mr. Fechner. - Obviously, pure coincidence, free will and unpredictable decisions are not determined processes. All terms denote the same phenomenon. Now where do we find such undetermined processes in physics?"

"The result of the decoherence of individual quanta is completely random and not determined. We cannot predict what real state they will take on. As we have seen in this course, quanta show primary consciousness. This reveals free will in the process of decoherence."

Professor Allman nods contentedly and turns to Elisabeth Delacroix, "How do you judge the larger units of consciousness, human consciousness for example? Do larger units of consciousness also possess free will?"

Professor Delacroix smiles: "I recall the behavioural traits by which we recognise consciousness. These behavioural characteristics include one with the formulation 'unpredictable, intentional behaviour'. The unpredictability, the unpredictability that is not predictable is one of the characteristics of consciousness. The undetermined decision is the prerequisite for such behaviour. Therefore consciousness already by definition possesses free will. If it weren't, we couldn't talk about consciousness, we'd have to call it zombiehood."

The auditorium and Professor Allman must laugh. The latter confirms and sums up: "You hit the nail on the head! That consciousness exists instead of zombie existence was proven during the course. - The free will of consciousness and the splitting of space-time universes in connection with conscious decisions ensure that for a consciousness the future is not fixed.

The greying philosopher Dr. Maupertius comes to life: **"The one who lets himself drift instead of taking his life actively and with conscious decisions into his own hands, his future is stuck!**

"I am not a philosopher, Dr. Maupertius, but your reversal seems to me to be correct!" says Professor Allman meaningfully. "because the **future belongs to whoever actively pursues it**.

The demiurge (world builder) and the answer to the question of meaning

Dr. Aniane interrupts this little philosophical skirmish. "To talk to Plato: Where is the demiurge in your continuum, Professor Allman? I mean the world builder who created everything?"

"I believe that the world or the Continuum was not created, Dr. Aniane. The continuum and space-time create themselves through continuous evolution!"

"What am I to understand?"

"Let me explain my position to you step by step. - Let us start from the definition of the continuum. The continuum is 'everything that exists'. It follows: The demiurge is part of the continuum and, if he is the creator of the continuum, he must first have created himself. The question that therefore needs to be answered is: How can something be created from nothing? - With this question one recognizes the back-reference: 'to create oneself'. - Let's ask the mathematician Dr. Anaximenes if she knows structures that create themselves.

"Of course there are such structures! The most famous example is the fractal with the name 'Apfelmännchen'. It results from the successive application of a very simple formula."

Professor Allman creates a projection of the Mandelbrot-Set in the middle of the hall and within easy reach of everyone.

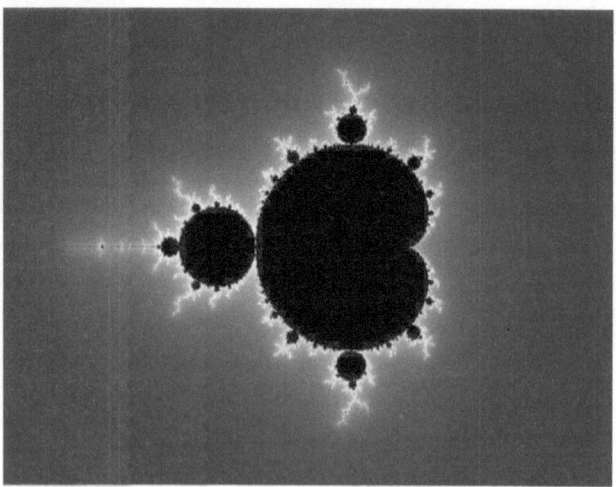

Figure 25: Famous fractal (male apple)

"Thank you, Professor Allman, if you could now show a detail of his edge?"

The physics professor complies with the request.

Figure 26: Detail of the edge of the Mandelbrot-Set enlarged fifty thousand times

"And what if you now show the incredibly simple formula that makes such a complex marvel practically come out of nothing by itself?"

Professor Allman first makes the participants marvel at the artworks and then projects the simple formula that leads to their creation.

$$Z \rightarrow Z^2 + c$$

The red-haired mathematician explains: "You can't tell from the formula what's in it. To get the image, start with the value $Z=0$, then calculate the right part, where c is a constant number. The result is one pixel. More pixels can be obtained by taking the result as the new value Z and recalculating the formula. You only need to repeat this many thousands of times and you will get more and more and finer details of the picture. There will always be new and surprising details of similar shapes. Thus an incredible complex work of art is created from a simple formula."

"I realize that the formula is very simple and the picture is very complex. But you need such a formula first. How can this come out of nowhere?" Dr. Aniane objects.

Professor Allman takes the answer. "That's no problem either. Every computer user has heard the term bit for the smallest unit of information. A bit can have one of two patterns, e.g. O or 1, and another possibility is nothing or something. - For the argumentation that follows, the new insights that you have gained in this course come into play. - Information consists of the parts pattern and meaning. But meaning only occurs together with consciousness. Consciousness is also a type of energy. Therefore a 'something' can be seen as the presence of elemental consciousness and energy, while a 'nothing' is the absence of both. **So if anything exists at all, it will be a combination of nothing and something.**"

Professor Allman emphasized the last sentence. Then he pauses for a few seconds to let his words take effect before continuing: "From an information technology point of view, these are bit combinations. For example, a byte is a bit combination consisting of eight bits. A byte, as you know for sure, means a letter or a character. - From the point of view of the continuum and vacuum theory, bit combinations are mergers of the smallest units of consciousness into larger ones.

The more they merge, the more complex the consciousness becomes. For the formula that brings forth the Mandelbrot-Set in its splendour, it does not even require a comprehensive fusion of these smallest units of consciousness. A merger of seven bytes or 56 bits is sufficient".

"But...", Dr. Aniane's head is spinning as he raises an objection. "...so space and time cannot be created!

Professor Allman smiles knowingly: "For a room in the manner of our viewing room, you only need one metric, that is a formula that assigns a distance to two points in space. The formula for the metric need not be larger than the formula for the Mandelbrot set. **Even a very small unit of consciousness can create a space of vision by making a metric become conscious.** - The other question of how consciousness makes time come into being has been discussed in the context of space-time.

"But... Dr. Aniane is unwell. He slides back and forth on his seat. "„... Surely all the glory of our universe could not have been created like this."

"Every decision in the decoherence of the quanta and every decision of a more complex consciousness leads not only to a division of the universe according to the Many-worlds theory, but also to the fact that consciousness stores more information and increases in its extent. Deciding means thinking. Through thinking a process of development is kept going, which develops the continuum and with it the space-time universe more and more, until the whole fullness has been created and even more. Another term for development is evolution. So let me repeat my opening sentence:

The continuum and spacetime create themselves through continuous evolution.

Dr. Aniane is grabbing his head. "But what about the master builder who created everything?"

"Do you remember Occam's razor, Dr. Aniane? - No need to be a master builder. Its introduction would complicate the theory. And what makes a theory more complicated, we don't need, we just cut it away with a razor."

Dr. Aniane is breathing heavily. "But then what's the meaning of life?"

"Have you ever thought that meaning is just another word for meaning? - It is the consciousness that gives meaning or sense. - Consciousness develops through thinking and decisions. So the point is that you live, think and decide consciously, and thereby advance the evolution of the world. To use a modified version of it: **The conscious path is the goal!** "

Dr. Aniane finds it difficult to process the new thoughts quickly. He gasps, "But what happens to us when we die?"

Immortality

"The question of what happens to human consciousness after the physical death of the body, considering the vacuum theory, can certainly be answered by someone other than a physicist. Professor Allman feels a little exhausted, but satisfied with what has been achieved so far. "Who dares say anything about the final topic of our seminar?"

Dr. Maupertius smiles wisely: "I think that as a philosopher I have the task to answer metaphysical questions.

"Then I beg you, Dr Maupertius, take this task from me!"

"Ladies and gentlemen, I believe we have gained profound new insights into our being in the last few days. I would like to incorporate these insights into my argumentation. - Consciousness, we have learned, is a function of the vacuum. The brain has the task of a receiver of consciousness. Consciousness does not need a brain for its existence. Physical death is a function of the spacetime universe and not of vacuum. Therefore physical death cannot harm the consciousness. It continues to exist in a vacuum. I could even imagine that after physical death it is freed from spatio-temporal restrictions and thus sees more clearly. Due to the static structure of four-dimensional space-time, consciousness can now simultaneously recognize space-time objects in their entire space-time extension, i.e. the complete film strip and not just individual images in seemingly temporal sequence. - I think I can even merge the concept of soul with that of consciousness by the following definition: A soul is the most highly motivated and energetically charged unit of consciousness. - So let me conclude by answering a question that has been discussed for thousands of years with one sentence: **People have a soul and it is immortal.** "

Professor Allman takes the floor again: "Thank you for your reference to philosophy, Dr Maupertius." Then he turns to all participants. "I need only add to the words of our esteemed colleague that all the knowledge you have gained here in the course is not based on revelation or esotericism, but solely on the scientific method. At the end of the course I may let you go home with the following request." This lights up in the middle of the room.

> Take an active part in the evolution!
>
> Develop the world through your thoughts and conscious decisions and do not be afraid of what will happen after the physical end, because man has an immortal consciousness.

When Professor Allman ends, the participants remain moved and calm for a short moment, then long lasting applause begins.

An informative week of training has come to an end.

Glossary

Decoherence	With the measurement of quanta, the phenomenon of decoherence takes place: a previously closed quantum system interacts with its environment. The measured quanta assume a real state, which they seem to choose for themselves. Before decoherence, the quanta were in an unreal state.
Determinism	Determinism is a philosophical concept. He assumes that all events take place in accordance with established laws and that they are completely determined by these laws.
Focus	The focus is on the element of reality that is perceived next by a consciousness.
Information	Information is a pattern of matter or a form of energy that has a certain meaning for an observer.
Interlocking	The entanglement is a phenomenon of quantum physics. Two or more entangled particles of a system are so closely related, even if they are spatially arbitrarily far apart, that the measurement of one particle directly influences the other. Considered by

	Einstein as a 'spooky remote effect'.
Length contraction	The contraction of length is a phenomenon of special relativity. For an observer, objects are shorter the faster they move relative to him.
Metrics	A metric is a mathematical function that assigns a value to two elements of a room, which can be understood as the distance between the two elements.
Occam's razor	Ockham's razor is an important decision criterion of science with the following statement: If there are several theories that explain the same facts, the simplest one is to choose.
Photons	Figuratively speaking, photons are something like "light particles".
polarization of light	Polarization is the alignment of the oscillation plane of light waves.
Quantum	A quantum is an energy packet of a certain size that shows quantum behaviour. Atomic and subatomic particles show quantum behaviour.
Spacetime	In the theory of relativity, space and time are united to form a uniform four-dimensional structure called space-time.
Spatio-temporal expansion	Spatio-temporal expansion means that an object contains all its

	eigenstates.
Time dilation	Time dilation is a phenomenon of the theory of relativity. It says that a watch that moves relative to an observer seems to run slower from the observer's point of view, and so does time itself.

References

Popularscientific books

Churchland, Paul M.: *Die Seelenmaschine. Eine philosophische Reise ins Gehirn*. Spektrum Akademischer Verlag Heidelberg, Berlin, 2001. Churchland vertritt die These, dass Denken, Fühlen und Selbstbewusstsein als reine Gehirnfunktion erklärt werden kann.

Davis, Paul: *Der Plan Gottes. Die Rätsel unserer Existenz und die Wissenschaft*. Insel Verlag, 1996.

Faulstich, Joachim: *Das Innere Land. Bewusstseinsreisen zwischen Leben und Tod*. Knaur, München 2006.

Laszlo, Ervin: *Holos die Welt der neuen Wissenschaften*. Verlag Via Nova, 2002. Eine Einführung in die grundlegenden Konzepte der Realität so, wie sie von Laszlo gesehen wird.

Lazlo, Ervin: *Zu Hause im Universum. Eine neue Vision der Wirklichkeit*. Allegria, 2005.

Pickover, Clifford A.: *Die Mathematik und das Göttliche*. Spektrum Akademischer Verlag Heidelberg, Berlin, 2003. Das Buch schärft auf amüsante Weise den Blick für die mathematische Struktur der Realität. Auch für Nichtmathematiker geeignet.

Rae, Alastair I.M.: *Quantenphysik: Illusion oder Realität?* Philipp Reclam jun. Stuttgart, 1996. Der britische Physiker erläutert die wichtigsten Punkte der Quantentheorie, diskutiert Probleme und deutet Lösungsmöglichkeiten an.

Tipler, Frank J.: *Die Physik der Unsterblichkeit. Moderne Kosmologie, Gott und die Auferstehung der Toten*. R. Piper GmbH & Co. KG, München 1994. Eine physikalische

Theorie, welche beansprucht die Versöhnung von Naturwissenschaft und Religion herbeizuführen. Der Autor sieht die Theologie als Spezialgebiet der Physik.

Zeilinger, Anton: *Einsteins Spuk. Teleportation und weitere Mysterien der Quantenphysik.* Wilhelm Goldmann Verlag, 2007. Zeilinger beschreibt, wie Teleportation funktioniert. Auch für Nichtphysiker nachvollziehbar!

Interesting articles

Buser, Pierre: *„Bewusstsein bei Tieren"*, in: Spektrum der Wissenschaft Spezial, Heft 1/2004.

Junge, J., D. Palu, F. Schön: *„Die scheinbare Welt"*, in: Welt der Wunder, Heft 2/2007. Wie das Gehirn die reale Welt simuliert.

Rees, Sir Martin: *„Ist das Leben eine Simulation?"*, in: Welt der Wunder, Heft 10/2007.

Ripota, Peter: *„Das Universum hat ein Bewusstsein!"*, in: P.M. Peter Moosleitners Magazin, September 2003.

Ripota, Peter: *„Wissenschaft kontra Religion: Schöpfungsmythen"*, in: P.M. Peter Moosleitners Magazin, April 2007.

Roth, Gerhard: *„Gleichtakt im Neuronennetz"*, in: Gehirn & Geist, Heft 1/2002. Der Autor versucht Bewusstsein als reine Gehirnfunktion zu erklären.

Schön, F., F. Meyer-Postell: *„Gibt es ein Bewusstsein außerhalb des Gehirns?"*, in: Welt der Wunder, Heft 8/2006.

Vaas, Rüdiger: *„Zeit ist nur eine Illusion"*, in: Bild der Wissenschaft, Heft 1/2008.

List of figures

Figure 1: Wegener's theory of displacement 19
Figure 2: The palaeobiogeographical distribution areas of Cynognathus, Mesosaurus, Glossopteris and Lystrosaurus shown here allow the reconstruction of the original continent 21
Figure 3: Ernest Rutherford at the time of his spreading experiment 25
Figure 4: Double-slit experiment 29
Figure 5: Interference bands on the photo plate 29
Figure 6: Plato 36
Figure 7: Continuum, Vacuum and Space-Time Universe (RZU) 40
Figure 8: Double-slit experiment in different variants 42
Figure 9: Two photon clusters without interference 47
Figure 10: Gravitational lensing effect. At each of the apparent locations a quasar can be seen. The gravitational field consists of a galaxy cluster. The earth is the focal point. Picture: Horst Frank. 51
Figure 11: Two-photon experiment for 'spooky long-distance effect'; the light source S generates photon pairs with the property that their polarization is always rectangular; PBS=polarizing beam splitter 53
Figure 12: Interlocked dice always show seven eyes after the throw. 56
Figure 13: Mathematical model for the continuum (=continuum model) 67
Figure 14: Atom model (not to scale) 73
Figure 15: Components of Information 77
Figure 16: Information as part of consciousness 80
Figure 17: How human consciousness was attempted to be

explained in the 17th century 84

Figure 18: Body temperature control system 87

Figure 19: Do common chimpanzees (Pan troglodytes) possess consciousness?   90

Figure 20: The shot-put pendulum illustrates a very short interaction with exchange of energy and information, Photo: Dominique Toussaint, GNU license see Appendix 97

Figure 21: Hyronimus Bosch; The Flight to Heaven; around 1500; Location: Doge's Palace in Venice   105

Figure 22: Minkowski diagram illustrating time dilation and length contraction. For the observer in A the moving clock in B seems to run slower. The length OD determined by him for the moving scale OE
is shorter.   114

Figure 23: Second generation NAVSTAR satellite for positioning   117

Figure 24: Temporal development and spatio-temporal expansion   118

Figure 25: Famous fractal (male apple)   126

Figure 26: Fifty thousand times enlarged detail of the edge of the Mandelbrot-Set   127

Keyword index

alpha particles 24
artificial intelligence 86
Awareness 7, 9, 64f., 80ff., 85ff., 91f., 94ff., 98ff., 105, 107ff., 113, 115, 120ff., 128ff., 133, 136f.
Back reference 126
Behavioural characteristics 86, 91ff., 124
Bell 45, 52, 57
Body temperature 87ff.
Brain waves 64f., 102
Brain 83, 99, 101f., 108, 120, 131, 136f.
central nervous system 88
Chimpanzees 87, 89, 91
cognitive awareness 85
Continental drift 20, 22, 27
Continuum model 99, 107
Continuum 39f., 68f., 125f., 129f.
Control system 87ff.
Creation myth 15f.
Credibility 9, 13, 18, 22, 27
curd 74
Death 9, 22, 92, 101f., 106, 130f., 136
Decoherence 47, 119, 122, 124, 129, 133
Demiurg 125f.
Descartes 83
Determinism 57f., 133
DeWitt 122
Disturbance 88
Double-slit experiment 27f., 30ff., 41, 49f., 52, 63, 93
Einstein 9, 18, 51, 53, 71, 78, 113, 137
Elementary block 76
Elementary particles 74f.
Encounter with Death 101
Ether Wind 112
ether 112
Everetts 122
Evolution 125, 130, 132
falsify 14f., 62
Formula 78, 124, 126ff.
free will 9, 123ff.
fundamental building block 110
Gallup 89
GPS satellite 116
Hypothesis 13f., 16ff., 20, 22f., 25ff., 30ff., 34, 60ff.
Illusion 111, 120, 123, 136f.
Immortality 9, 130f., 136
Information memory 62f., 70, 72, 78, 81, 96, 109
Information 39, 45, 48ff., 52, 57f., 63, 76ff., 80, 93ff., 98f., 109f., 128f., 133
Interaction 47, 71f., 96, 98ff., 107ff., 121, 133
Interference 28, 42f., 46, 48, 52, 63, 93
Interlocking 54, 135
Length contraction 113, 116ff., 133
Life after death 106
Magnetic resonance imaging 83
Mandelbrot-Set 126, 129
Many-worlds-theory 121f., 129
Master builder 130
mathematical model 30, 31, 32, 107
Meaning 11, 14, 53, 76ff., 85, 113, 116, 128, 130
Metaphysics 11f., 18, 33, 83
metric free 99, 109
Metric 68ff., 72, 96, 109, 129, 134
Michelson interferometer 112
Michelson-Morley Experiment 111
Minkowski

Diagram 114
muons 117
Near-death experience 101
Neuroscience 86
non-local connection 69, 108
non-local 69., 99, 107ff.
Nothing 89, 100, 126ff.
Nuclear 23f., 26ff., 74
Observer 113, 115f., 120
Occam's razors 14, 17, 61, 108, 122, 130, 134
Operation 101f., 104ff., 108
Pam Reynolds 101ff.
Parapsychology 18
path-time diagram 115, 118
Phenomena 20ff., 26f., 30ff., 34f., 39, 45, 47f., 57f., 60ff., 83, 98f., 101, 133
Photon 41, 43ff., 51ff., 62f., 69, 94, 100, 107, 119, 121f., 134
physical death 130f.
Plato's Allegory of the Cave 95
Plato 35f., 73, 125
Polarization 54, 94
Polarization 54ff., 134
pre-scientific 18, 22
Prediction 14f., 17f., 22, 30ff., 64f., 124
primary consciousness 86, 89, 91f., 94ff., 100, 124
Pseudoscience 12, 16
Quant 9f., 27, 30, 32, 39f., 45, 47ff., 53f., 57, 61ff., 69ff., 78, 92ff., 98ff., 109, 113, 119, 122ff., 129, 133f., 136f.
Quantum mechanics 27, 30, 32, 71
Quantum theory 113, 123, 136
Quasar 50ff.
Random 55, 57f., 121, 124
Reality 9, 11f., 14ff., 20, 23, 26f., 31ff., 38f., 51, 58, 60, 71ff., 77, 79f., 85f., 88, 91, 93, 95, 100, 111, 122, 133, 136
Reality 9, 33, 58, 116, 136
Relativity theory 51, 71, 113ff., 120, 134f.
Religion 12f., 15f.
Rutherford 24, 26f.
RZU 39ff., 45f., 48f., 60f., 99, 107f., 118
Samples 41, 65, 77ff., 85, 98, 128
Scattering test 24, 27
Science 7, 9ff., 18, 22, 26, 32ff., 35, 39f., 60, 83, 86, 92, 104f., 108, 136f.
Self-confidence 136
Simultaneity 113
Sinn 9, 100, 125, 130
Skeptic Movement 104
Something 82, 120, 128
Soul 58, 131
Space and time 45, 50, 52f., 58, 60f., 66f., 69, 111, 113f., 118, 129, 134
Space-time universe 45, 50, 52, 58, 61ff., 69, 72, 77, 95, 110
Space-time 51, 71, 117ff., 123, 125, 129ff., 134
spatio-temporal expansion 119, 135
Speed of light 58, 69, 112, 115, 117
Spirit 9, 13, 83, 113, 137
spooky remote effect 53, 58, 62, 69, 107, 135
Stain test 89, 92
Teleportation 53, 137
temporal development 118ff.
Theory 12ff., 20, 22, 26f., 30ff., 45f., 50, 57ff., 63f., 66, 70f., 99
Time dilation 113, 116ff., 135
Time is an illusion 120, 123
Time 13, 20, 39f., 44f., 47, 50, 52f., 58, 60ff., 65ff., 72, 77, 83, 89, 95, 104, 108, 110f., 113ff., 118, 120f., 123, 137
topological space 70, 72, 96, 109
transpersonal consciousness 64, 99f.
Two-photon experiment 58, 62, 94, 107, 121f.

141

unexplained phenomena 20, 22f., 27, 35, 62f., 111f.
Unit of consciousness 109, 124, 129, 131
Vacuum energy 96
Vacuum 16, 39ff., 45ff., 53, 58, 60f., 63ff., 68ff., 78, 81, 95f., 98ff., 108f., 113, 131
verify 14f.
Viewing room 111, 119, 129
Wave-particle duality 27, 30, 32f., 48
Wegener 20ff., 27, 60
World builder 125
Zeilinger 53, 137

Appendix

Some specially marked images are licensed under GNU. The following license terms apply to these images:

GNU Free Documentation License

Version 1.2, November 2002

Copyright (C) 2000,2001,2002 Free Software Foundation, Inc.
51 Franklin St, Fifth Floor, Boston, MA 02110-1301 USA
Everyone is permitted to copy and distribute verbatim copies
of this license document, but changing it is not allowed.

0. PREAMBLE

The purpose of this License is to make a manual, textbook, or other functional and useful document "free" in the sense of freedom: to assure everyone the effective freedom to copy and redistribute it, with or without modifying it, either commercially or noncommercially. Secondarily, this License preserves for the author and publisher a way to get credit for their work, while not being considered responsible for modifications made by others.
This License is a kind of "copyleft", which means that derivative works of the document must themselves be free in the same sense. It complements the GNU General Public License, which is a copyleft license designed for free software.
We have designed this License in order to use it for manuals for free software, because free software needs free documentation: a free program should come with manuals providing the same freedoms that the software does. But this License is not limited to software manuals; it can be used for any textual work, regardless of subject matter or whether it is published as a printed book. We recommend this License principally for works whose purpose is instruction or reference.

1. APPLICABILITY AND DEFINITIONS

This License applies to any manual or other work, in any medium, that contains a notice placed by the copyright holder saying it can be distributed under the terms of this License. Such a notice grants a world-wide, royalty-free license, unlimited in duration, to use that work under the conditions stated herein. The "Document", below, refers to any such manual or work. Any member of the public is a licensee, and is addressed as "you". You accept the license if you copy, modify or distribute the work in a way requiring permission under copyright law.
A "Modified Version" of the Document means any work containing the Document or a portion of it, either copied verbatim, or with modifications and/or translated into another language.
A "Secondary Section" is a named appendix or a front-matter section of the Document that deals exclusively with the relationship of the publishers or authors of the Document to the Document's overall subject (or to related matters) and contains nothing that could fall directly within that overall subject. (Thus, if the Document is in part a textbook of mathematics, a Secondary Section may not explain any mathematics.) The relationship could be a matter of historical connection with the subject or with related matters, or of legal, commercial, philosophical, ethical or political position regarding them.
The "Invariant Sections" are certain Secondary Sections whose titles are designated, as being those of Invariant Sections, in the notice that says that the Document is released under this License. If a section does not fit the above definition of Secondary then it is not allowed to be designated as Invariant. The Document may contain zero Invariant Sections. If the Document does not identify any Invariant Sections then there are none.
The "Cover Texts" are certain short passages of text that are listed, as Front-Cover Texts or Back-Cover Texts, in the notice that says that the Document is released under this License. A Front-Cover Text may be at most 5 words, and a Back-Cover Text may be at most 25 words.
A "Transparent" copy of the Document means a machine-readable copy, represented in a format whose specification is available to the general public, that is suitable for revising the document straightforwardly with generic text editors or (for images composed of pixels) generic paint programs or (for drawings) some widely available drawing editor, and that is suitable for input to text formatters or for automatic translation to a variety of formats suitable for input to text formatters. A copy made in an otherwise Transparent file format whose markup, or absence of markup, has been arranged to thwart or discourage subsequent modification by readers is not Transparent. An image format is not Transparent if used for any substantial amount of text. A copy that is not "Transparent" is called "Opaque".
Examples of suitable formats for Transparent copies include plain ASCII without markup, Texinfo input format, LaTeX input format, SGML or XML using a publicly available DTD, and standard-conforming simple HTML, PostScript or PDF designed for human modification. Examples of transparent image formats include PNG, XCF and JPG. Opaque formats include proprietary formats that can be read and edited only by proprietary word processors, SGML or XML for

which the DTD and/or processing tools are not generally available, and the machine-generated HTML, PostScript or PDF produced by some word processors for output purposes only.

The "Title Page" means, for a printed book, the title page itself, plus such following pages as are needed to hold, legibly, the material this License requires to appear in the title page. For works in formats which do not have any title page as such, "Title Page" means the text near the most prominent appearance of the work's title, preceding the beginning of the body of the text.

A section "Entitled XYZ" means a named subunit of the Document whose title either is precisely XYZ or contains XYZ in parentheses following text that translates XYZ in another language. (Here XYZ stands for a specific section name mentioned below, such as "Acknowledgements", "Dedications", "Endorsements", or "History".) To "Preserve the Title" of such a section when you modify the Document means that it remains a section "Entitled XYZ" according to this definition.

The Document may include Warranty Disclaimers next to the notice which states that this License applies to the Document. These Warranty Disclaimers are considered to be included by reference in this License, but only as regards disclaiming warranties: any other implication that these Warranty Disclaimers may have is void and has no effect on the meaning of this License.

2. VERBATIM COPYING

You may copy and distribute the Document in any medium, either commercially or noncommercially, provided that this License, the copyright notices, and the license notice saying this License applies to the Document are reproduced in all copies, and that you add no other conditions whatsoever to those of this License. You may not use technical measures to obstruct or control the reading or further copying of the copies you make or distribute. However, you may accept compensation in exchange for copies. If you distribute a large enough number of copies you must also follow the conditions in section 3.

You may also lend copies, under the same conditions stated above, and you may publicly display copies.

3. COPYING IN QUANTITY

If you publish printed copies (or copies in media that commonly have printed covers) of the Document, numbering more than 100, and the Document's license notice requires Cover Texts, you must enclose the copies in covers that carry, clearly and legibly, all these Cover Texts: Front-Cover Texts on the front cover, and Back-Cover Texts on the back cover. Both covers must also clearly and legibly identify you as the publisher of these copies. The front cover must present the full title with all words of the title equally prominent and visible. You may add other material on the covers in addition. Copying with changes limited to the covers, as long as they preserve the title of the Document and satisfy these conditions, can be treated as verbatim copying in other respects.

If the required texts for either cover are too voluminous to fit legibly, you should put the first ones listed (as many as fit reasonably) on the actual cover, and continue the rest onto adjacent pages.

If you publish or distribute Opaque copies of the Document numbering more than 100, you must either include a machine-readable Transparent copy along with each Opaque copy, or state in or with each Opaque copy a computer-network location from which the general network-using public has access to download using public-standard network protocols a complete Transparent copy of the Document, free of added material. If you use the latter option, you must take reasonably prudent steps, when you begin distribution of Opaque copies in quantity, to ensure that this Transparent copy will remain thus accessible at the stated location until at least one year after the last time you distribute an Opaque copy (directly or through your agents or retailers) of that edition to the public.

It is requested, but not required, that you contact the authors of the Document well before redistributing any large number of copies, to give them a chance to provide you with an updated version of the Document.

4. MODIFICATIONS

You may copy and distribute a Modified Version of the Document under the conditions of sections 2 and 3 above, provided that you release the Modified Version under precisely this License, with the Modified Version filling the role of the Document, thus licensing distribution and modification of the Modified Version to whoever possesses a copy of it. In addition, you must do these things in the Modified Version:

- **A.** Use in the Title Page (and on the covers, if any) a title distinct from that of the Document, and from those of previous versions (which should, if there were any, be listed in the History section of the Document). You may use the same title as a previous version if the original publisher of that version gives permission.

- **B.** List on the Title Page, as authors, one or more persons or entities responsible for authorship of the modifications in the Modified Version, together with at least five of the principal authors of the Document (all of its principal authors, if it has fewer than five), unless they release you from this requirement.

- **C.** State on the Title page the name of the publisher of the Modified Version, as the publisher.

- **D.** Preserve all the copyright notices of the Document.

- **E.** Add an appropriate copyright notice for your modifications adjacent to the other copyright notices.

- **F.** Include, immediately after the copyright notices, a license notice giving the public permission to use the Modified Version under the terms of this License, in the form shown in the Addendum below.

- **G.** Preserve in that license notice the full lists of Invariant Sections and required Cover Texts given in the Document's license notice.
- **H.** Include an unaltered copy of this License.
- **I.** Preserve the section Entitled "History", Preserve its Title, and add to it an item stating at least the title, year, new authors, and publisher of the Modified Version as given on the Title Page. If there is no section Entitled "History" in the Document, create one stating the title, year, authors, and publisher of the Document as given on its Title Page, then add an item describing the Modified Version as stated in the previous sentence.
- **J.** Preserve the network location, if any, given in the Document for public access to a Transparent copy of the Document, and likewise the network locations given in the Document for previous versions it was based on. These may be placed in the "History" section. You may omit a network location for a work that was published at least four years before the Document itself, or if the original publisher of the version it refers to gives permission.
- **K.** For any section Entitled "Acknowledgements" or "Dedications", Preserve the Title of the section, and preserve in the section all the substance and tone of each of the contributor acknowledgements and/or dedications given therein.
- **L.** Preserve all the Invariant Sections of the Document, unaltered in their text and in their titles. Section numbers or the equivalent are not considered part of the section titles.
- **M.** Delete any section Entitled "Endorsements". Such a section may not be included in the Modified Version.
- **N.** Do not retitle any existing section to be Entitled "Endorsements" or to conflict in title with any Invariant Section.
- **O.** Preserve any Warranty Disclaimers.

If the Modified Version includes new front-matter sections or appendices that qualify as Secondary Sections and contain no material copied from the Document, you may at your option designate some or all of these sections as invariant. To do this, add their titles to the list of Invariant Sections in the Modified Version's license notice. These titles must be distinct from any other section titles.

You may add a section Entitled "Endorsements", provided it contains nothing but endorsements of your Modified Version by various parties--for example, statements of peer review or that the text has been approved by an organization as the authoritative definition of a standard.

You may add a passage of up to five words as a Front-Cover Text, and a passage of up to 25 words as a Back-Cover Text, to the end of the list of Cover Texts in the Modified Version. Only one passage of Front-Cover Text and one of Back-Cover Text may be added by (or through arrangements made by) any one entity. If the Document already includes a cover text for the same cover, previously added by you or by arrangement made by the same entity you are acting on behalf of, you may not add another; but you may replace the old one, on explicit permission from the previous publisher that added the old one.

The author(s) and publisher(s) of the Document do not by this License give permission to use their names for publicity for or to assert or imply endorsement of any Modified Version.

5. COMBINING DOCUMENTS

You may combine the Document with other documents released under this License, under the terms defined in section 4 above for modified versions, provided that you include in the combination all of the Invariant Sections of all of the original documents, unmodified, and list them all as Invariant Sections of your combined work in its license notice, and that you preserve all their Warranty Disclaimers.

The combined work need only contain one copy of this License, and multiple identical Invariant Sections may be replaced with a single copy. If there are multiple Invariant Sections with the same name but different contents, make the title of each such section unique by adding at the end of it, in parentheses, the name of the original author or publisher of that section if known, or else a unique number. Make the same adjustment to the section titles in the list of Invariant Sections in the license notice of the combined work.

In the combination, you must combine any sections Entitled "History" in the various original documents, forming one section Entitled "History"; likewise combine any sections Entitled "Acknowledgements", and any sections Entitled "Dedications". You must delete all sections Entitled "Endorsements".

6. COLLECTIONS OF DOCUMENTS

You may make a collection consisting of the Document and other documents released under this License, and replace the individual copies of this License in the various documents with a single copy that is included in the collection, provided that you follow the rules of this License for verbatim copying of each of the documents in all other respects. You may extract a single document from such a collection, and distribute it individually under this License, provided you insert a copy of this License into the extracted document, and follow this License in all other respects regarding verbatim copying of that document.

7. AGGREGATION WITH INDEPENDENT WORKS

A compilation of the Document or its derivatives with other separate and independent documents or works, in or on a volume of a storage or distribution medium, is called an "aggregate" if the copyright resulting from the compilation is not used to limit the legal rights of the compilation's users beyond what the individual works permit. When the Document is included in an aggregate, this License does not apply to the other works in the aggregate which are not themselves derivative works of the Document.
If the Cover Text requirement of section 3 is applicable to these copies of the Document, then if the Document is less than one half of the entire aggregate, the Document's Cover Texts may be placed on covers that bracket the Document within the aggregate, or the electronic equivalent of covers if the Document is in electronic form. Otherwise they must appear on printed covers that bracket the whole aggregate.

8. TRANSLATION

Translation is considered a kind of modification, so you may distribute translations of the Document under the terms of section 4. Replacing Invariant Sections with translations requires special permission from their copyright holders, but you may include translations of some or all Invariant Sections in addition to the original versions of these Invariant Sections. You may include a translation of this License, and all the license notices in the Document, and any Warranty Disclaimers, provided that you also include the original English version of this License and the original versions of those notices and disclaimers. In case of a disagreement between the translation and the original version of this License or a notice or disclaimer, the original version will prevail.
If a section in the Document is Entitled "Acknowledgements", "Dedications", or "History", the requirement (section 4) to Preserve its Title (section 1) will typically require changing the actual title.

9. TERMINATION

You may not copy, modify, sublicense, or distribute the Document except as expressly provided for under this License. Any other attempt to copy, modify, sublicense or distribute the Document is void, and will automatically terminate your rights under this License. However, parties who have received copies, or rights, from you under this License will not have their licenses terminated so long as such parties remain in full compliance.

10. FUTURE REVISIONS OF THIS LICENSE

The Free Software Foundation may publish new, revised versions of the GNU Free Documentation License from time to time. Such new versions will be similar in spirit to the present version, but may differ in detail to address new problems or concerns. See http://www.gnu.org/copyleft/.
Each version of the License is given a distinguishing version number. If the Document specifies that a particular numbered version of this License "or any later version" applies to it, you have the option of following the terms and conditions either of that specified version or of any later version that has been published (not as a draft) by the Free Software Foundation. If the Document does not specify a version number of this License, you may choose any version ever published (not as a draft) by the Free Software Foundation.

ADDENDUM: How to use this License for your documents

To use this License in a document you have written, include a copy of the License in the document and put the following copyright and license notices just after the title page:
Copyright (c) YEAR YOUR NAME.

Permission is granted to copy, distribute and/or modify this document
under the terms of the GNU Free Documentation License, Version 1.2
or any later version published by the Free Software Foundation;
with no Invariant Sections, no Front-Cover Texts, and no Back-Cover Texts.
A copy of the license is included in the section entitled
"GNU Free Documentation License."

If you have Invariant Sections, Front-Cover Texts and Back-Cover Texts, replace the "with...Texts." line with this:
with the Invariant Sections being LIST THEIR TITLES, with the
Front-Cover Texts being LIST, and with the Back-Cover Texts being LIST.
If you have Invariant Sections without Cover Texts, or some other combination of the three, merge those two alternatives to suit the situation.

If your document contains nontrivial examples of program code, we recommend releasing these examples in parallel under your choice of free software license, such as the GNU General Public License, to permit their use in free software.

The nature of the physical world

The Gifford Lectures 1927

By Sir Arthur Eddington , Klaus-Dieter Sedlacek (Hrsg.)

In these lectures the author Eddington discusses some of the results of modern study of the physical world which give most food for philosophic thought. This will include new conceptions in science and also new knowledge. In both respects we are led to think of the material universe in a way very different from that prevailing at the classical physics.

This book is substantially the course of Gifford Lectures which the author Eddington delivered in the University of Edinburgh in January to March 1927. It treats of the philosophical outcome of the great changes of scientific thought. The theory of relativity and the quantum theory have led to strange new conceptions of the physical world; the progress of the principles of thermodynamics has wrought more gradual but no less profound change.

www.ingramcontent.com/pod-product-compliance
Lightning Source LLC
Chambersburg PA
CBHW031419210526
45464CB00005B/1952